Handwriting Book

- Put your counters on the house.
- Take turns to spin the spinner.
- Move to the nearest larger or smaller bear.
- The winner is the first to get to the biggest bear.

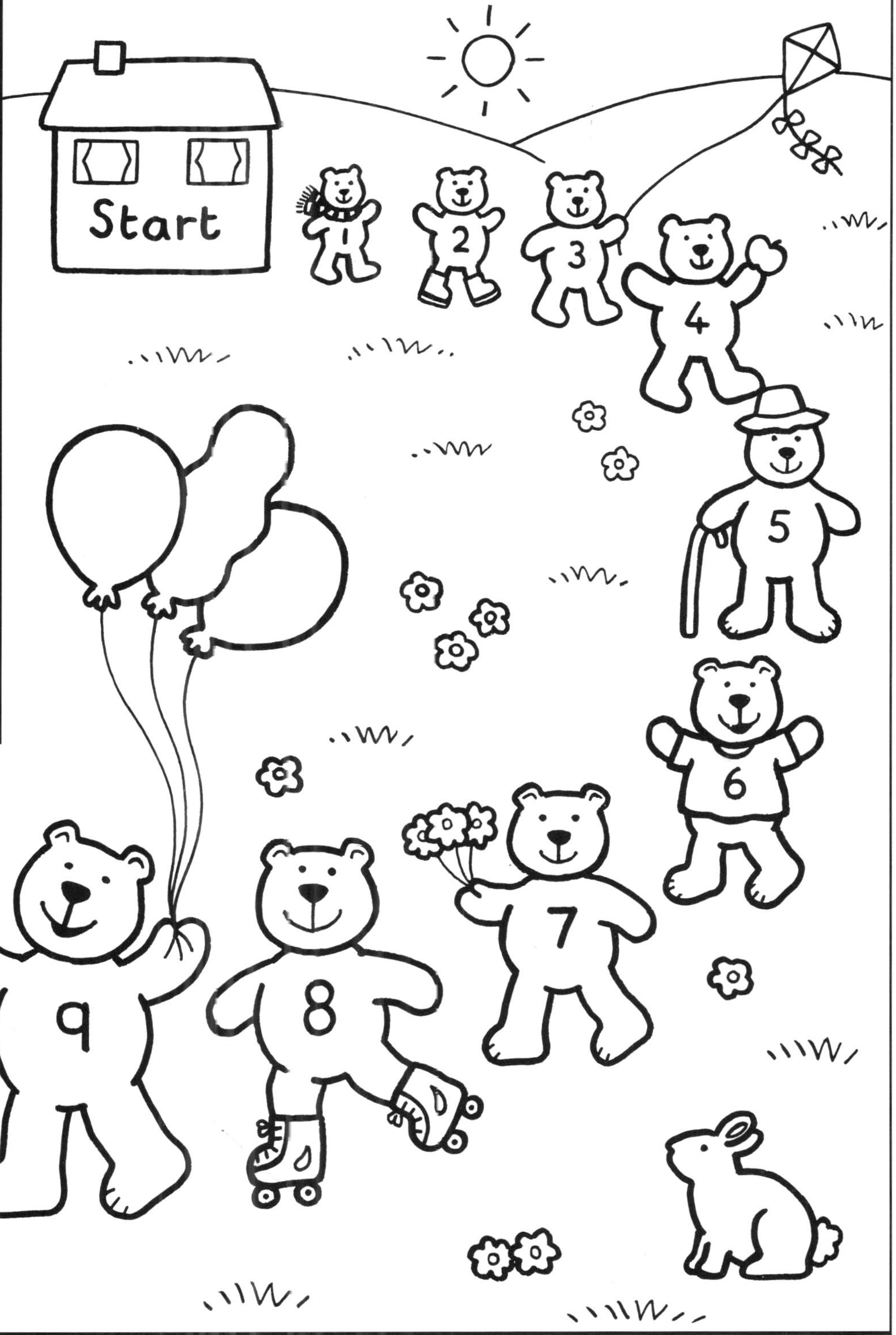

Name . . . . Class . . . .

1

I can write these numbers	Colour the face	Teacher check
1	☺ almost ☺ yes	
2	☺ ☺	
3	☺ ☺	
4	☺ ☺	
5	☺ ☺	
6	☺ ☺	
7	☺ ☺	
8	☺ ☺	
9	☺ ☺	
10	☺ ☺	

Trace

1 2 3 4 5

1 1 1 1

2 2 2 2

3 3 3 3

4 4 4 4

5 5 5 5

Copy the numbers

1 2 3 1 2 3 4 5

2	• Writing numbers to 5. Notes/date:

3

Finish the numbers.

| 4 | ● Writing numbers that go right.   Notes/date |

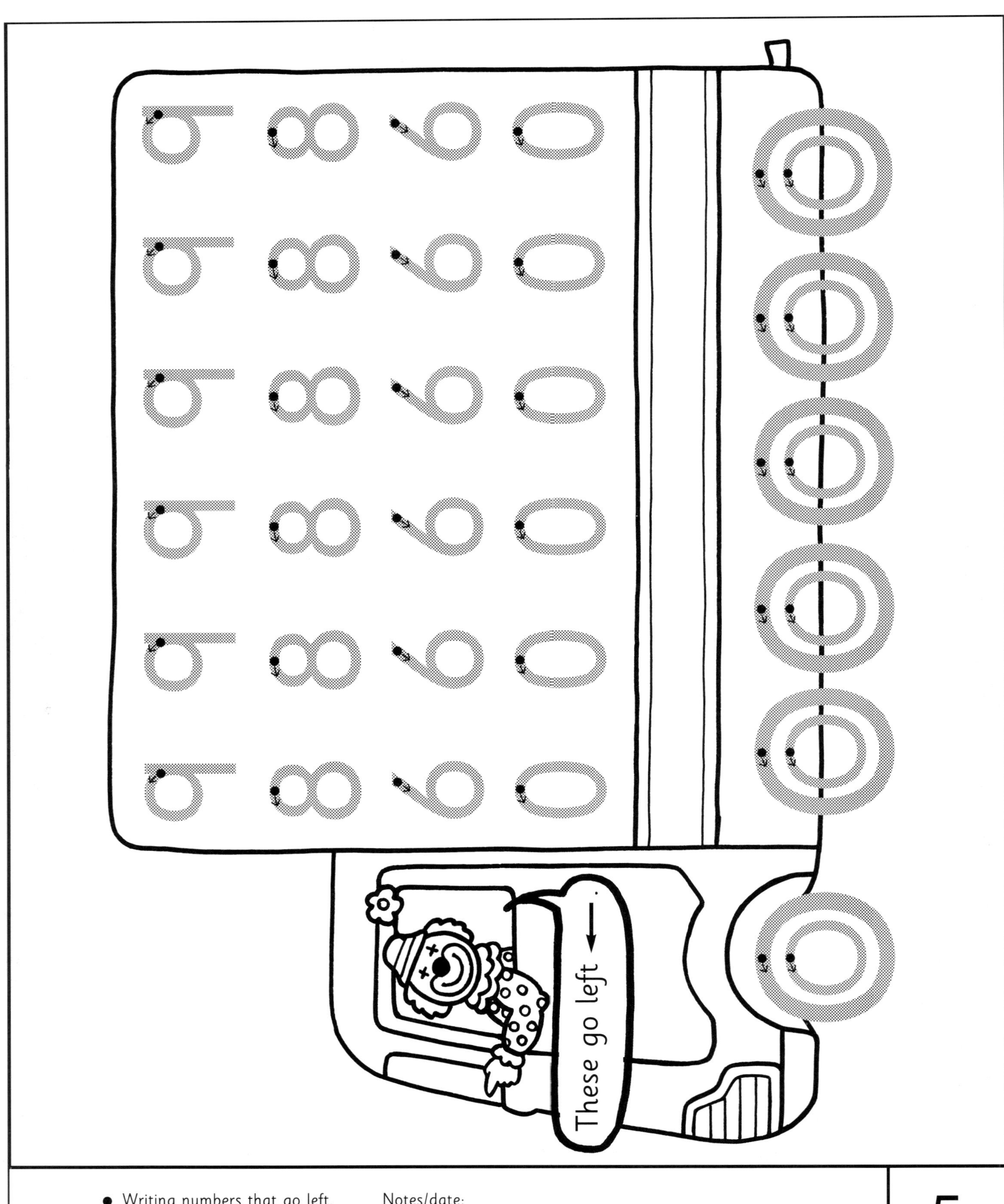

● Writing numbers that go left.   Notes/date:

5

0 1 2 3 4 5 6 7 8 9

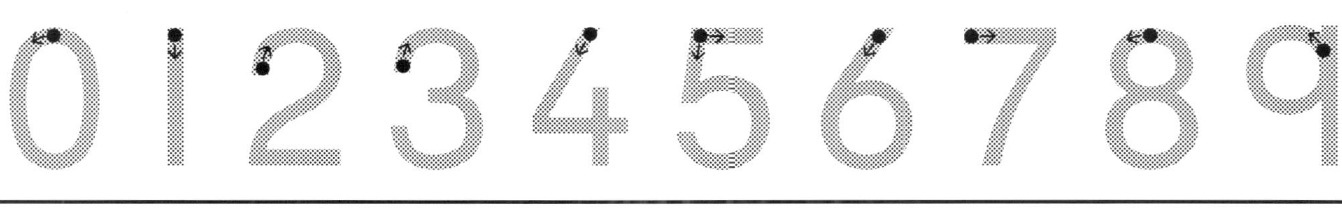

0 0 0 0 0   1 1 1 1 1

2 2 2 2   3 3 3 3

4 4 4 4   5 5 5 5

6 6 6 6   7 7 7 7

8 8 8 8   9 9 9 9

6

• Writing numbers 0–9.   Notes/date

Name . . . .   Class . . . .

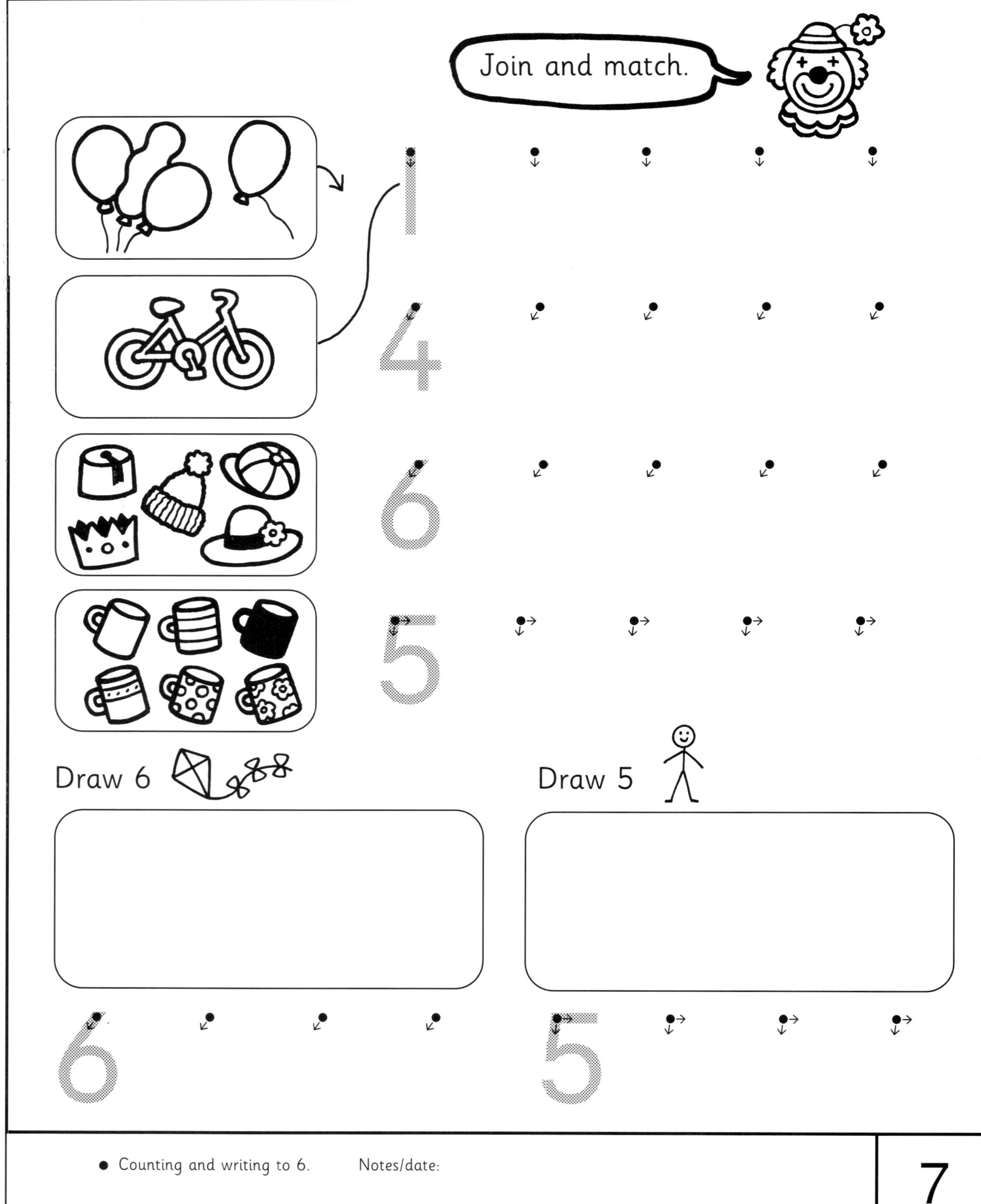

Join and match.

1
4
6
5

Draw 6

Draw 5

6

5

8	● Writing numbers 0–9. Notes/date:

● Counting and writing to ___ .   Notes/date:

q

Count and write.

I can write these numbers.

| 10 | • Counting and writing to ___ .   Notes/date: |

0 1 2 3 4 5 6 7 8 9 10

How many?

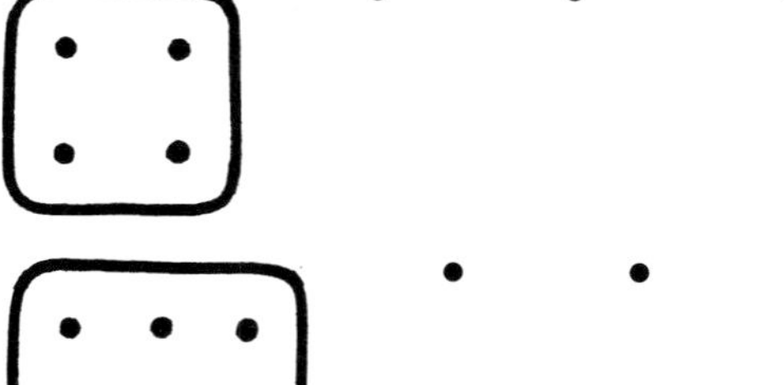

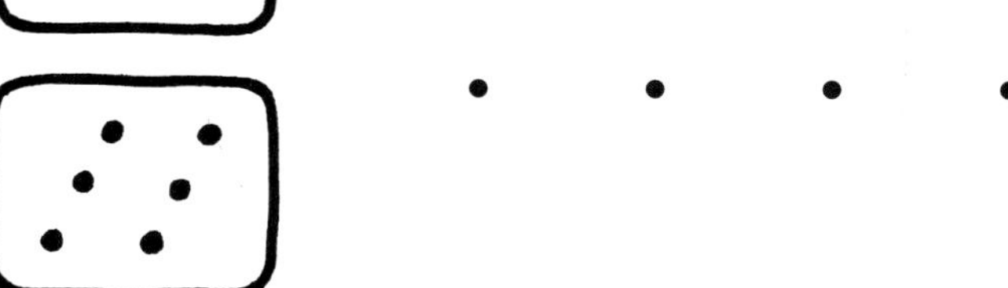

Draw

2 

7

8

9

2

7

8

9

Count and write.

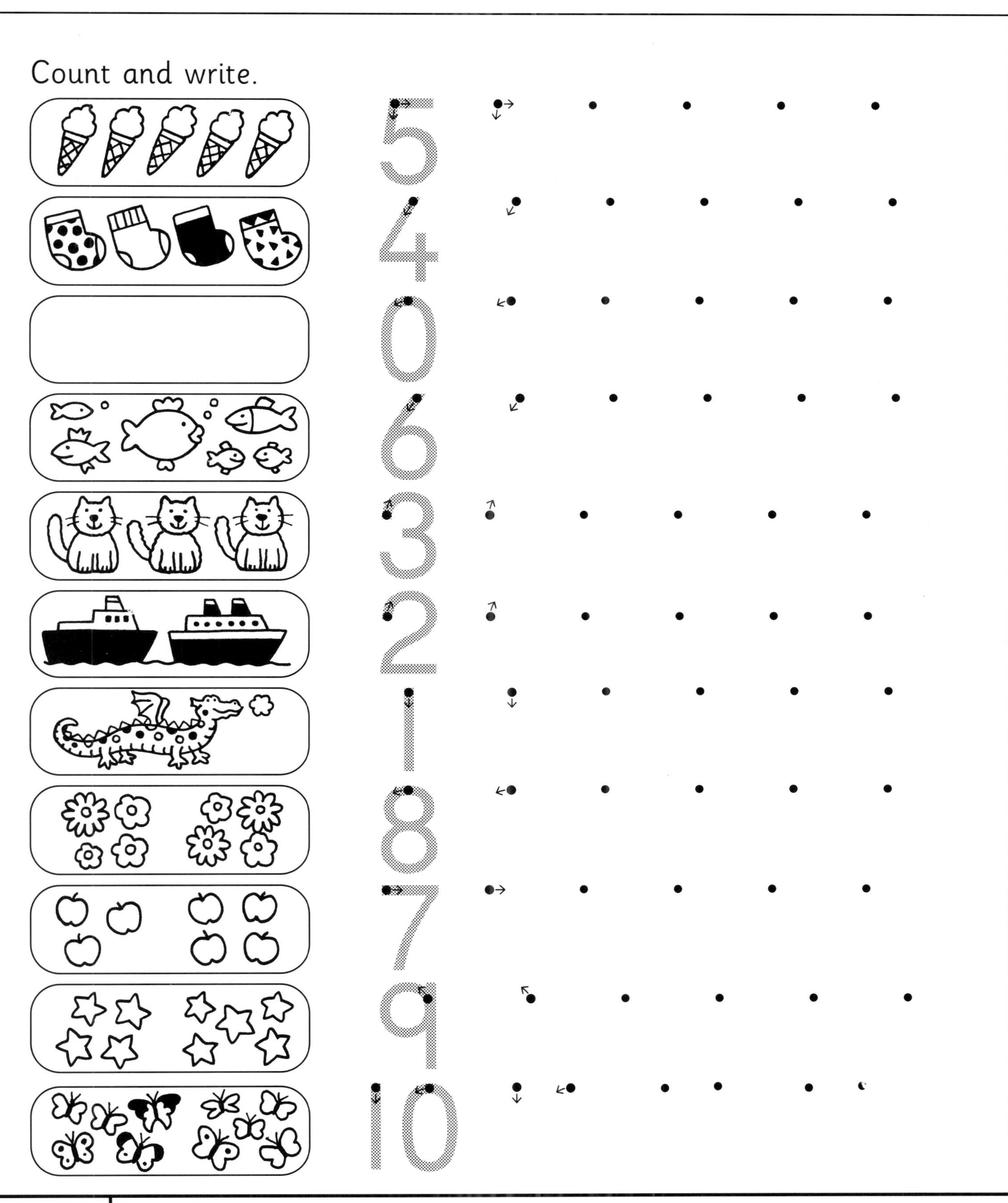

| 12 | ● Counting and writing to 0–10.   Notes/date: |

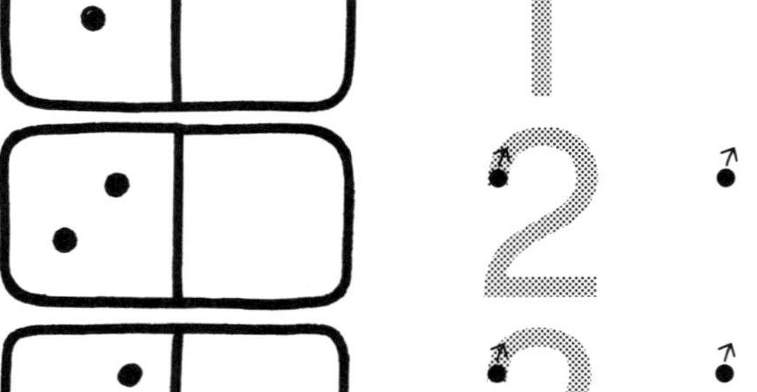

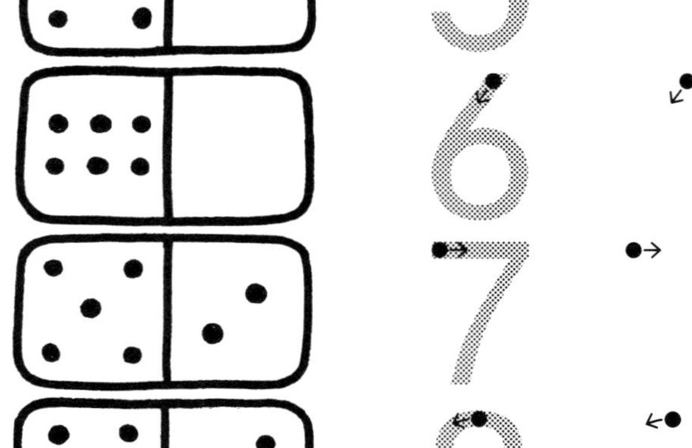

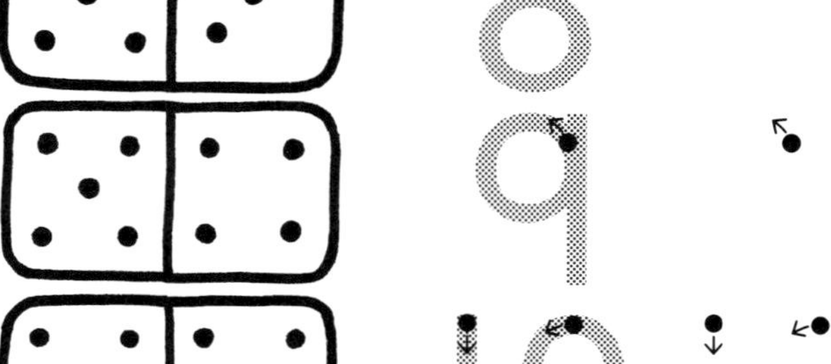

Draw dots.

• Counting and writing to 0–10.   Notes/date:

13

Name . . . . Class . . . .

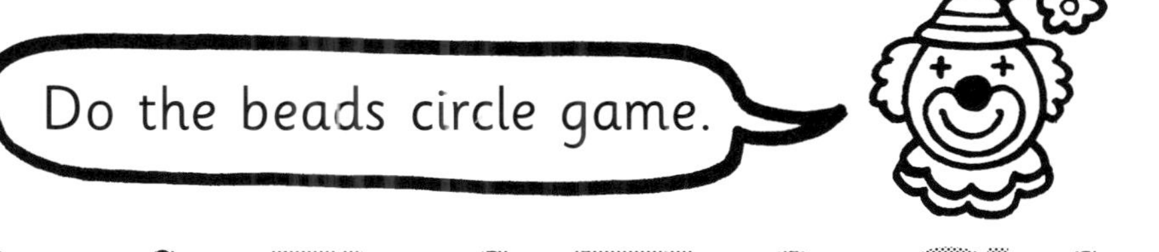

0 1 2 3 4 5 6 7 8 9 10

How many?

4

3

Draw

Draw

I can write these numbers.

14	• Counting and visualising numbers as 5 and a bit more.	Notes/date:

Name . . . . **Class** . . . .

I can write these numbers.

I can count.

Draw [ ]

Draw [ ]

I can ...

Notes/date:

15

Practice book I

Dragon's game

You need
- a partner
- a spotty dice 1–6
- 6 counters each
- a book each

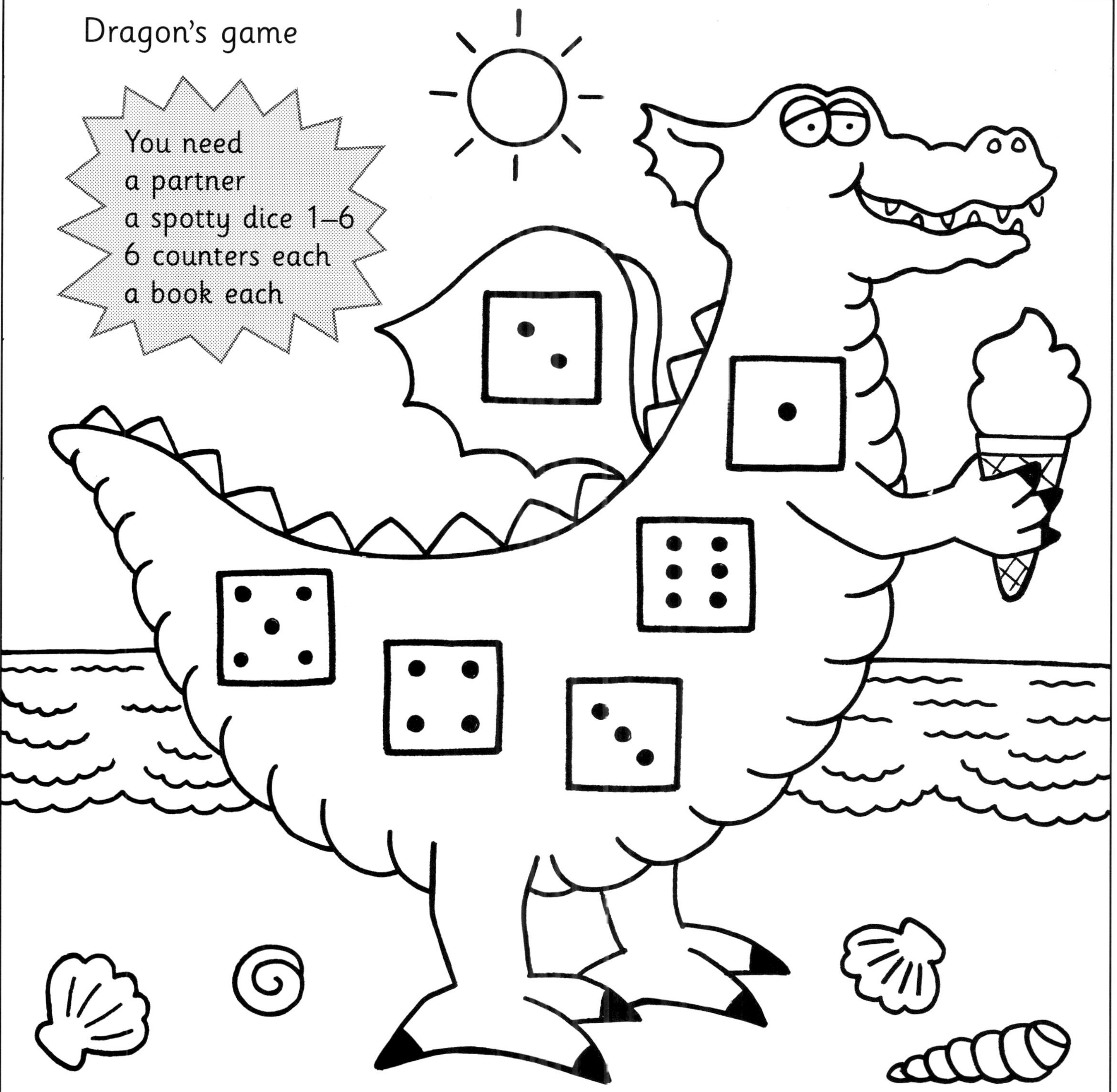

Name . . . . **Class** . . . .

Tally how many times you play games.

Bears bigger and bigger Handwriting book game	
Dragon's game	

I can count to

I am going to try to

1. --

2. --

I can   --

----

date ___________

Dragon's game

Rules
- Each use your own book.
- Take turns to throw the dice.
- Cover the number with a counter.
- The winner is the first to cover all their numbers.

Name . . . . Class . . . .

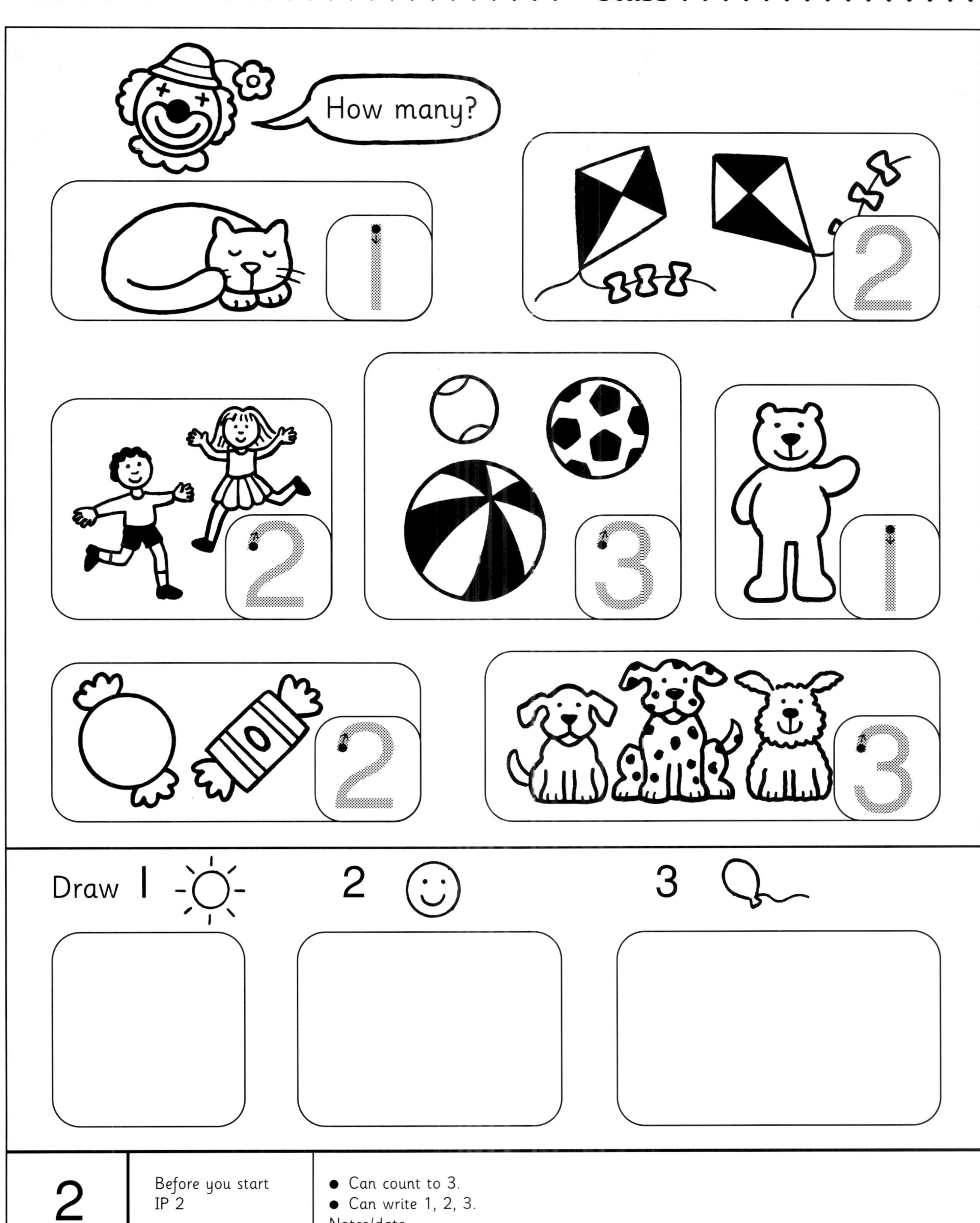
How many?
1
2
2
3
1
2
3
Draw 1   2   3
2   Before you start
     IP 2
● Can count to 3.
● Can write 1, 2, 3.
Notes/date:

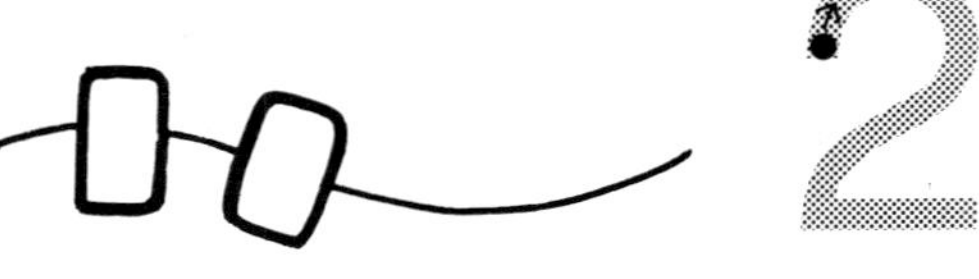

Choose how many

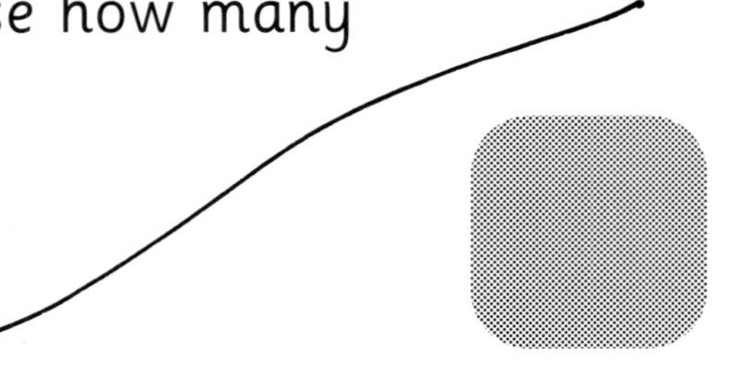

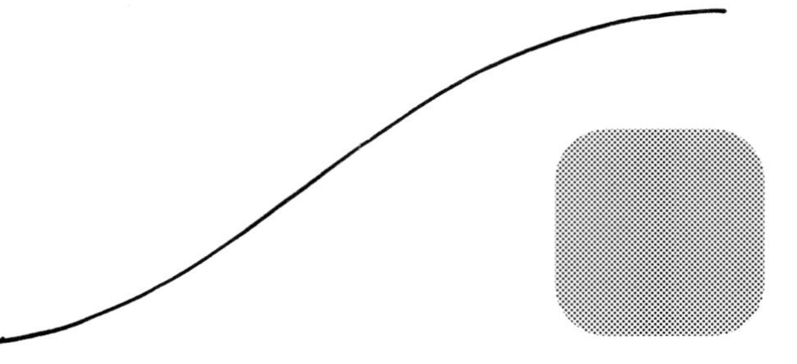

Join

- Can count to 3.
- Can write 1, 2, 3.

Notes/date:

Before you start
IP 14, 2
CG Beads

3

| 4 | Before you start IP 4 | ● Can join related items.
Notes/date: |

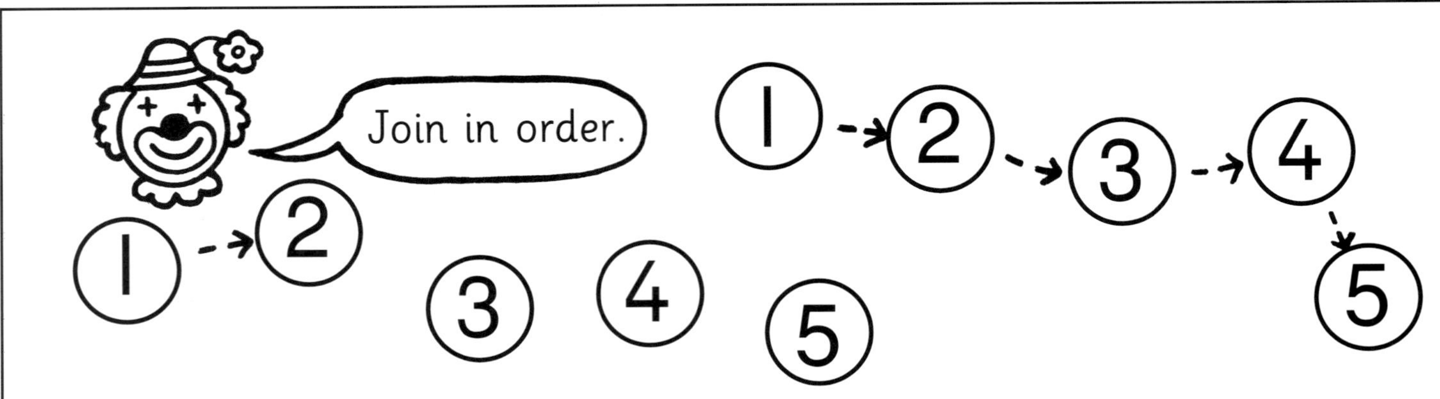

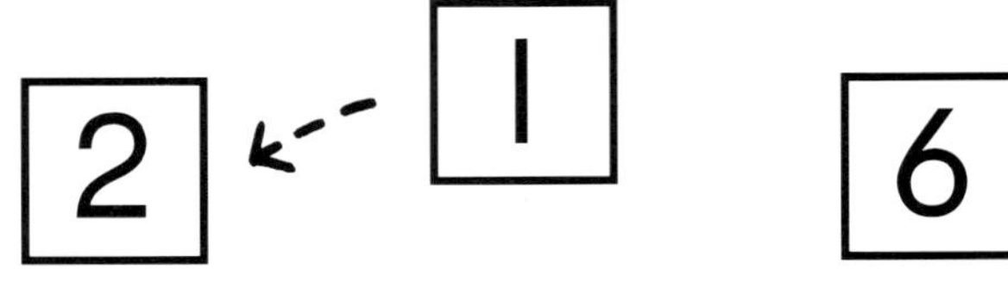

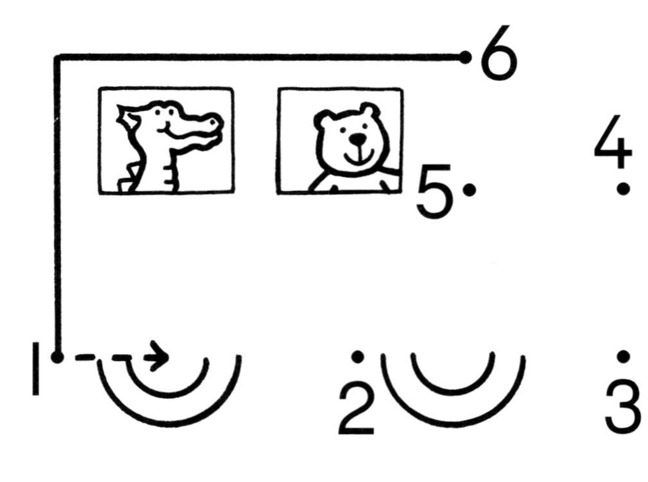

Fill in the missing numbers.

● Can order numbers to 6.
Notes/date:

Before you start
IP 2

5

4 4

How many ● ?

Tick ✓ sets of 4.

4 four 4 four

4 4 4 4 4 4 4 4

6	Before you start IP 4	● Can count to 4. Notes/date

5 5

How many ● ?

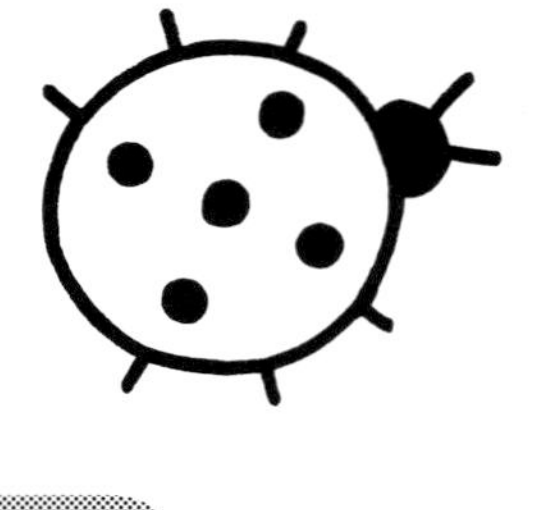

Tick ✓ sets of 5.

Draw
5 🍎

Draw
5 ✳

5 five 5 5 5 5 5

● Can count to 5.
Notes/date:

Before you start
IP 4

7

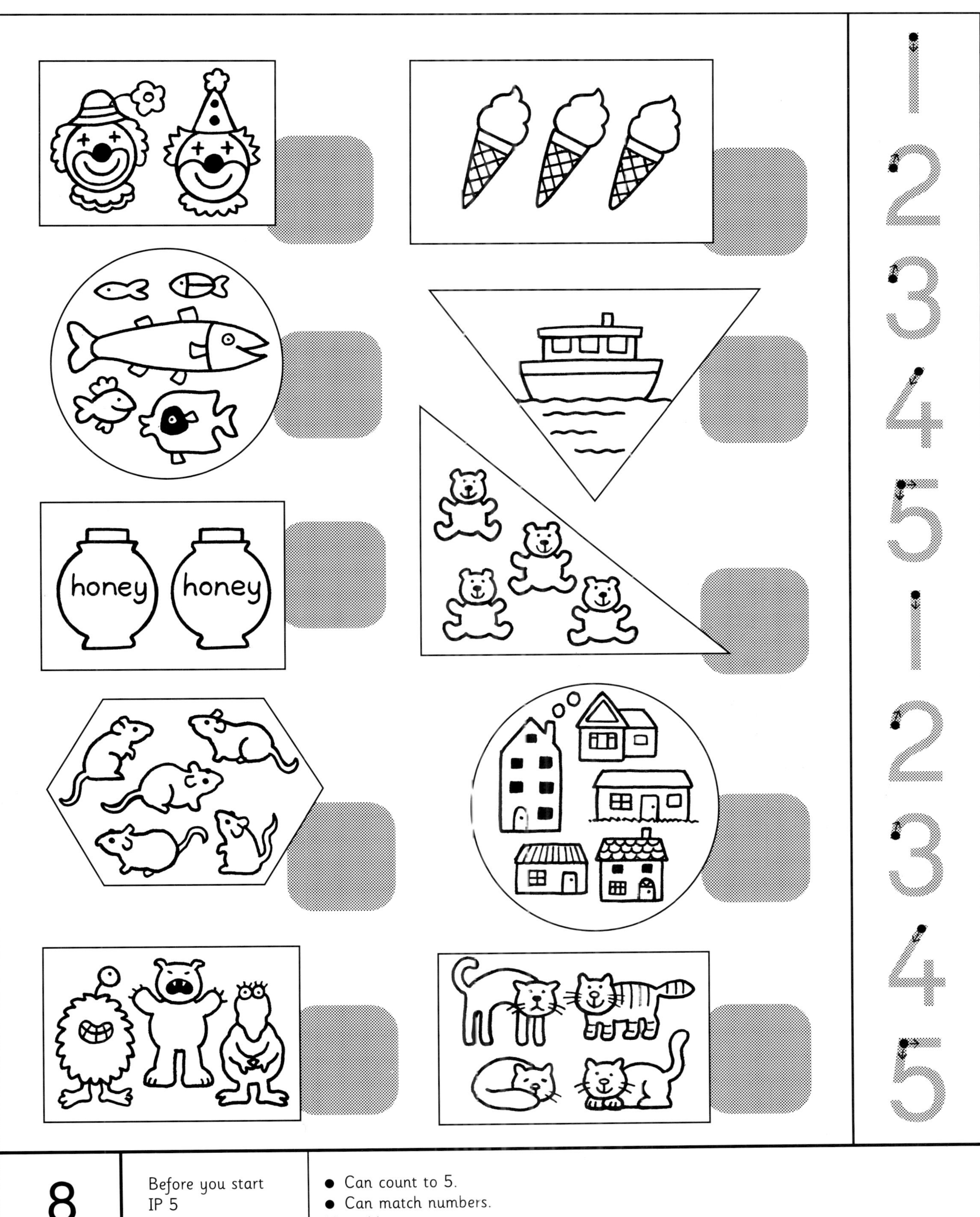

8

Before you start
IP 5

● Can count to 5.
● Can match numbers.
Notes/date:

Name . . . . **Class** . . . .

Draw 2 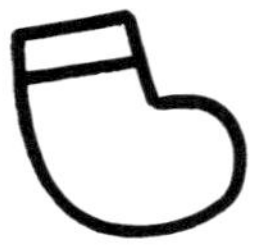	
Draw 1 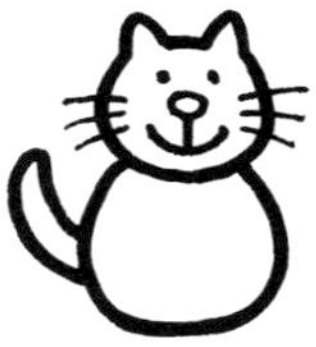	
Draw 3 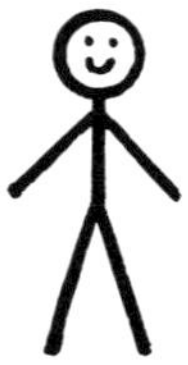	
Draw 5 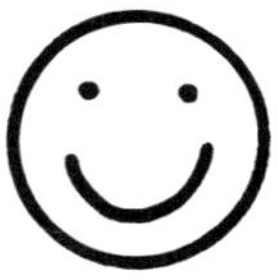	
Draw 4 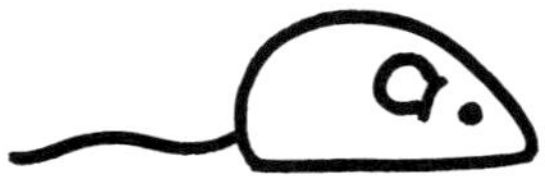	

• Can count to 5 / above. • Recognises numbers. Notes/date:	Before you start IP 5	9

Colour

How many?

| 10 | Before you start
IP 9 | ● Can count to 5.
● Recognises shapes.
Notes/date: |

Name . . . . Class . . . .

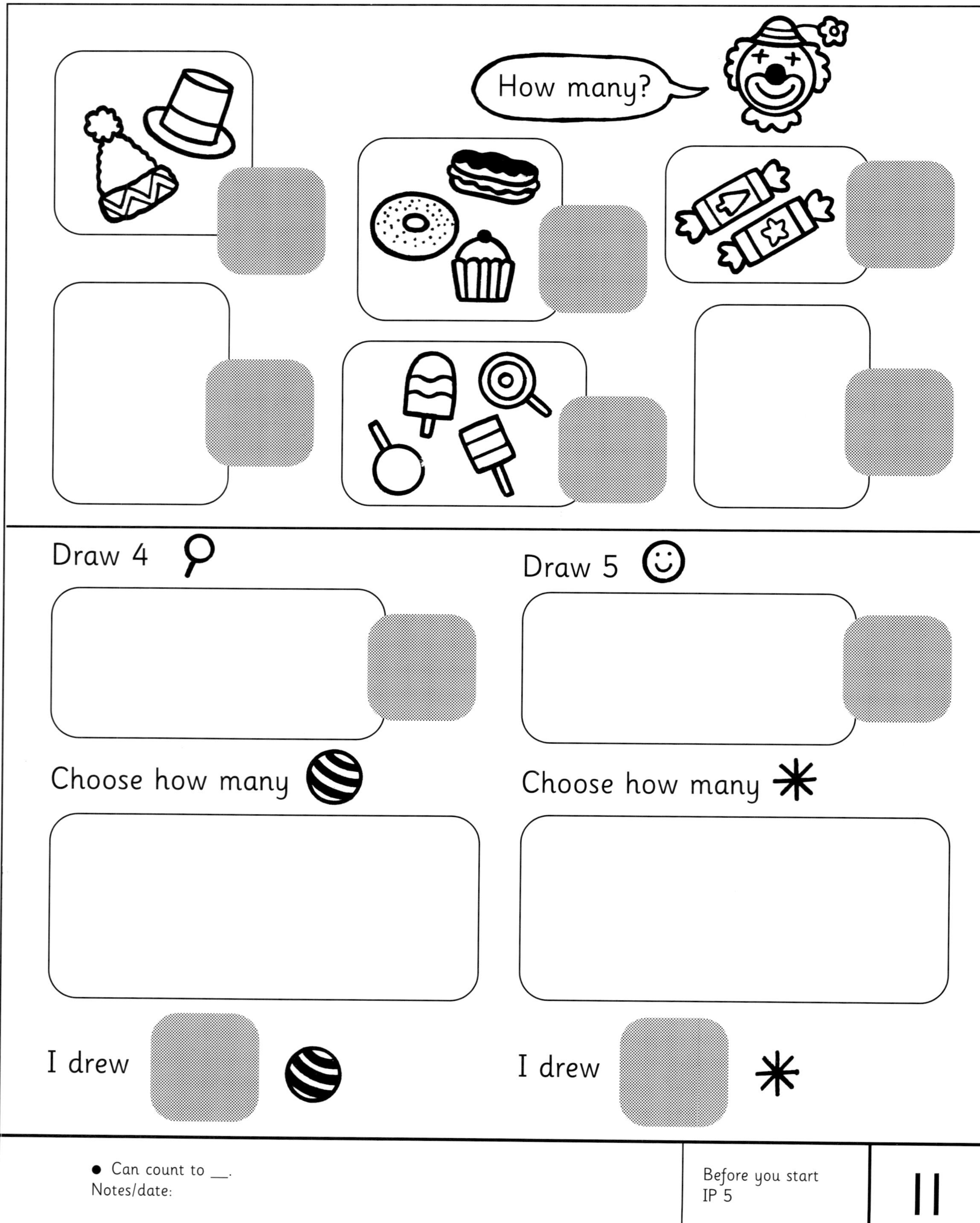
How many?

Draw 4
Draw 5
Choose how many
Choose how many
I drew
I drew

● Can count to __.
Notes/date:

Before you start
IP 5

11

Name . . . . Class . . . .

12 | Before you start IP 5 | ● Can use the language of 'more than'. Notes/date:

Draw ● to make 5.

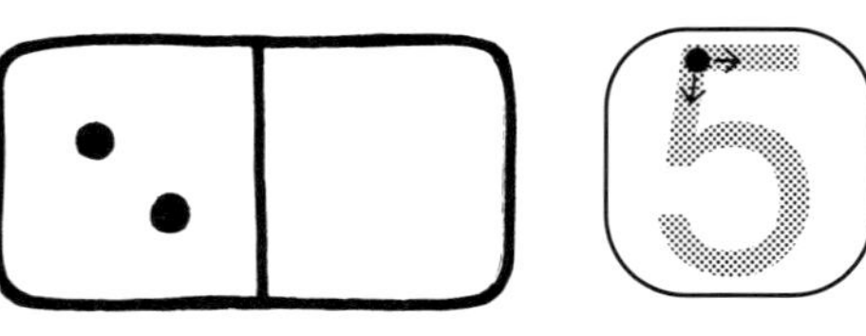 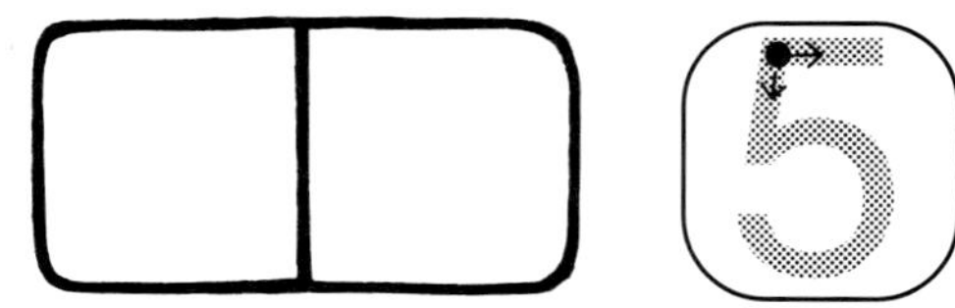

Draw ● to make 6.

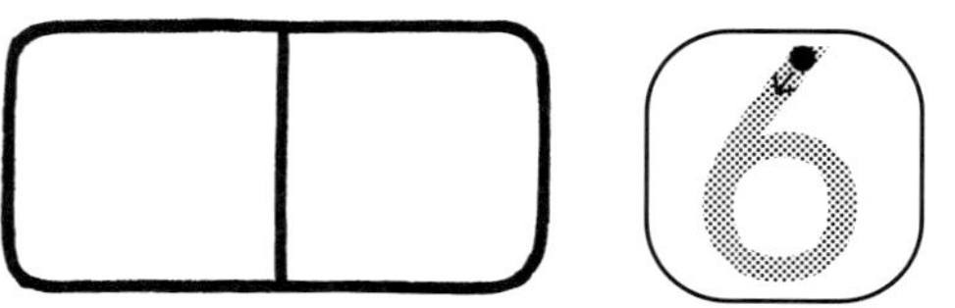 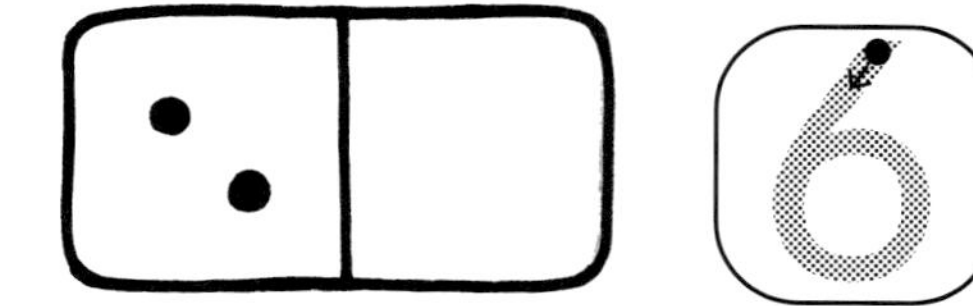

Draw ● to make 

 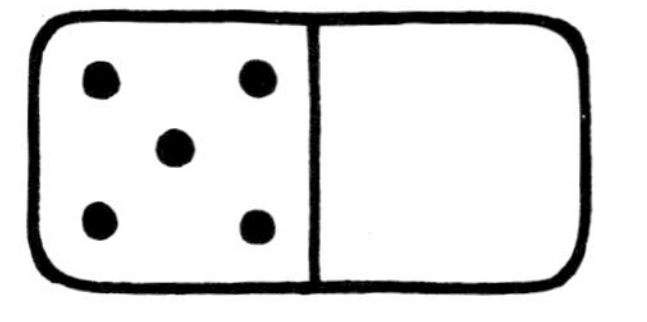 

● Can count to 6 / above.
● Can make up to 6 / above.
Notes/date:

Before you start
IP 2

13

Draw what you did.

14	Before you start IP 13 CG Mice in a hole	● Can record splitting of numbers. Notes/date:

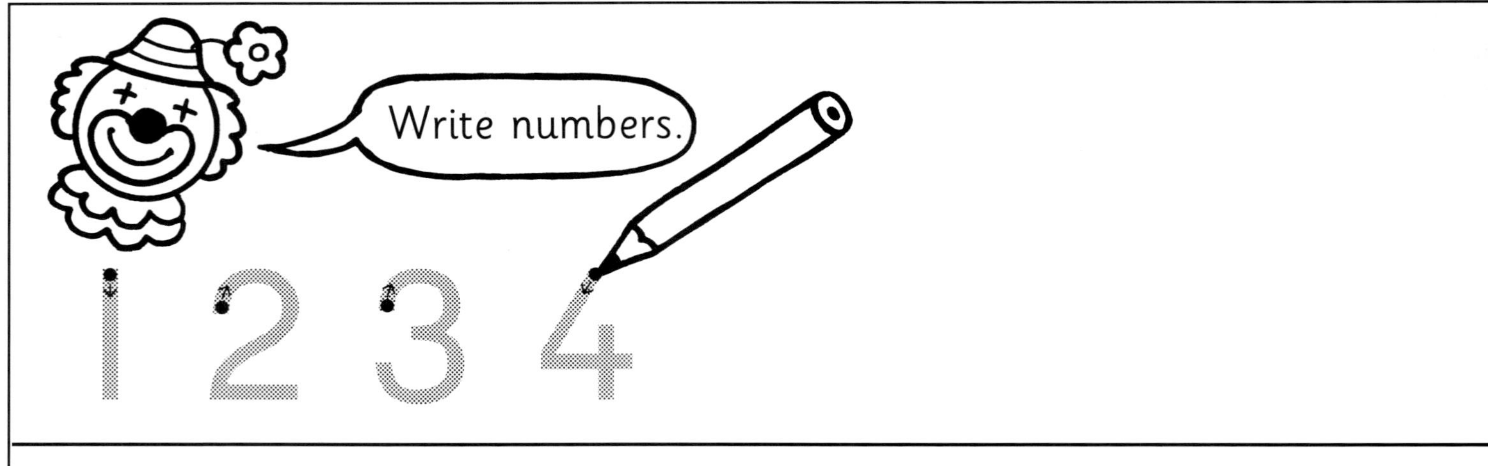

I can write numbers up to

Choose how many ●

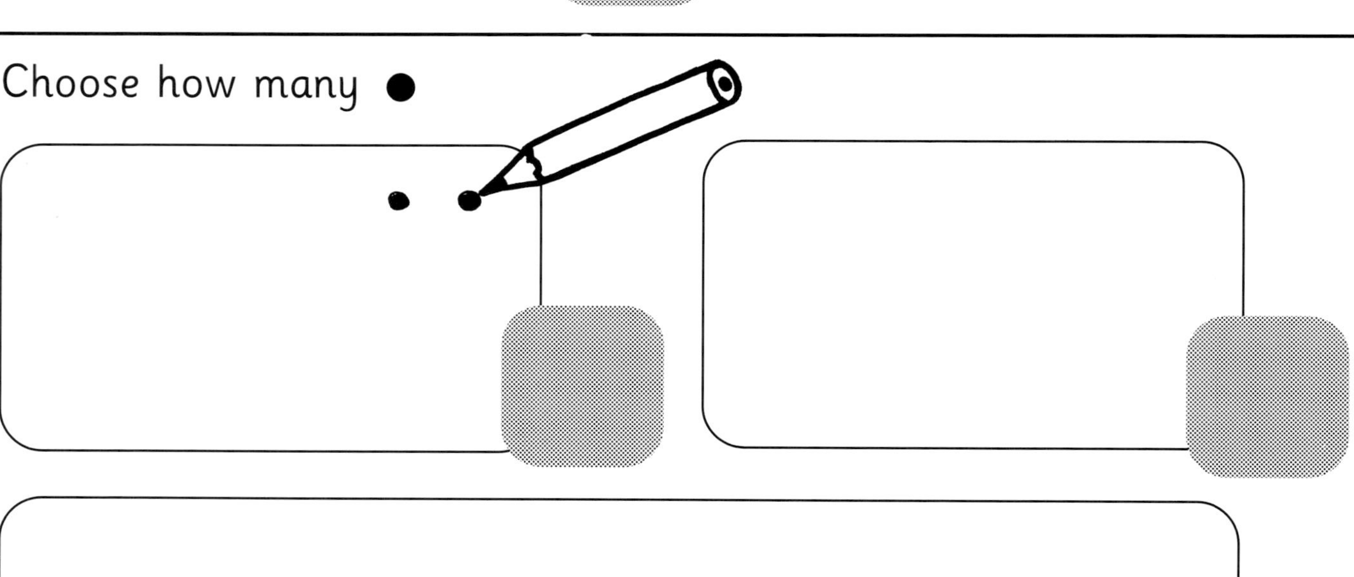

- Can write and say numbers to __.   Notes/date:
- Can count to __

For assessment discussion

15

First Skills in Numeracy 1

Practice Book 2

Balloon game

You need
- a partner
- a book each
- cubes
- a dice 1–6

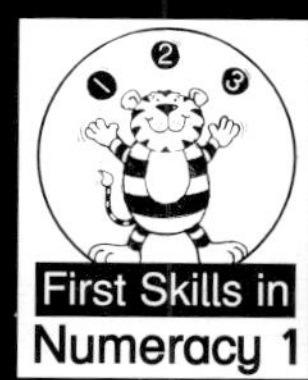

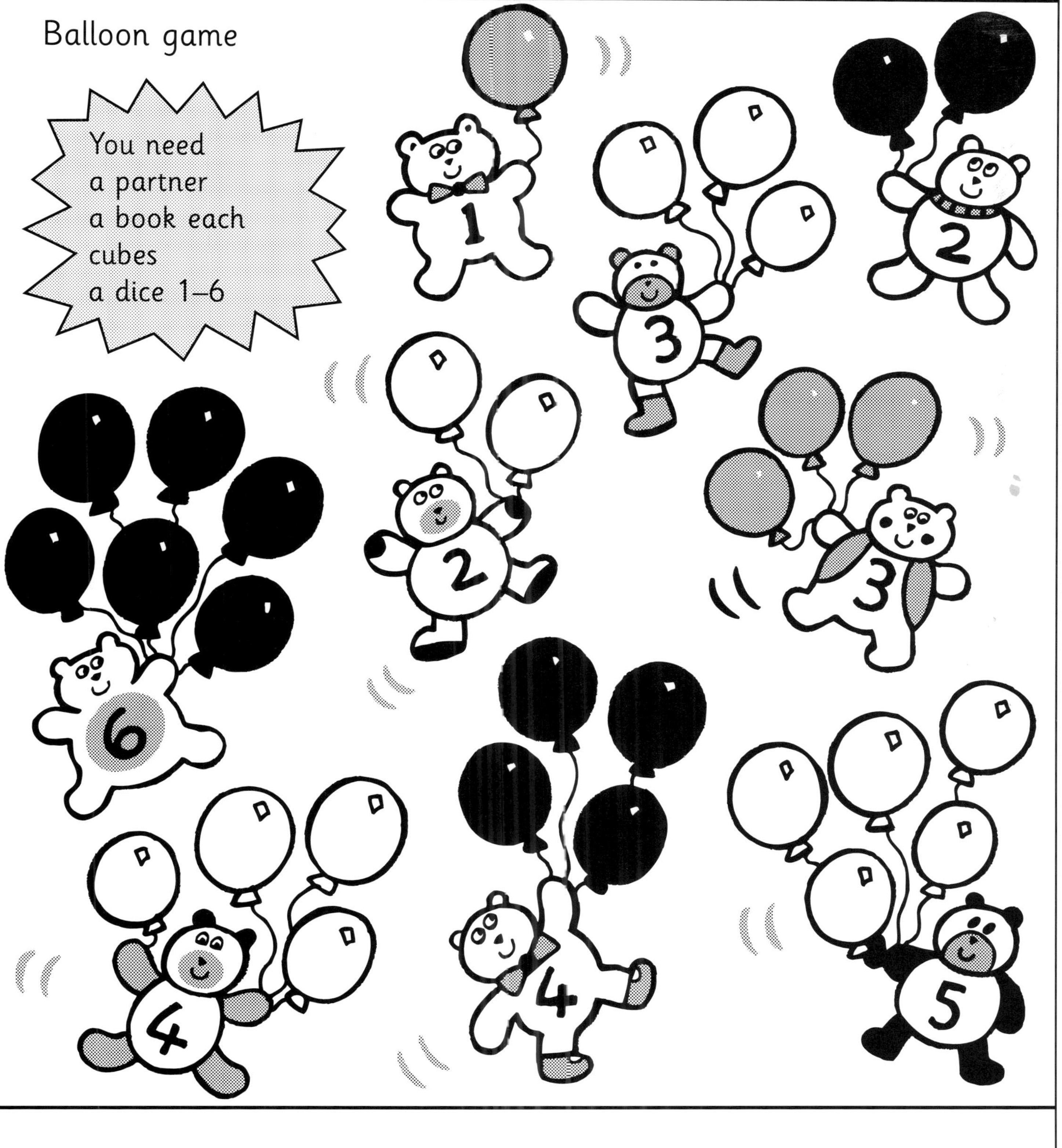

Name . . . . **Class** . . . .

Name . . . . Class . . . .

Tally how many times you play games. I I I /

Balloon game

I can count to

I am going to try to

   1. --

   2. --

I can   --

----

                date ___________

Balloon game

Game 1
- Take turns to throw the dice.
- Match the number to a teddy.
- Cover the teddy with a cube.
- The winner is the first to cover all their teddies.

Game 2
- Put a cube on each balloon.
- Take turns to throw the dice.
- Take away that number of cubes.
- The winner is the first to uncover all their balloons.

| 2 | Before you start
IP 1
CG Find the partner | ● Can count to 9.
Notes/date: |

● Can match numbers, objects and words, 0–10.
Notes/date:

Before you start
IP 2

3

Name . . . . **Class** . . . .

Draw the coins.

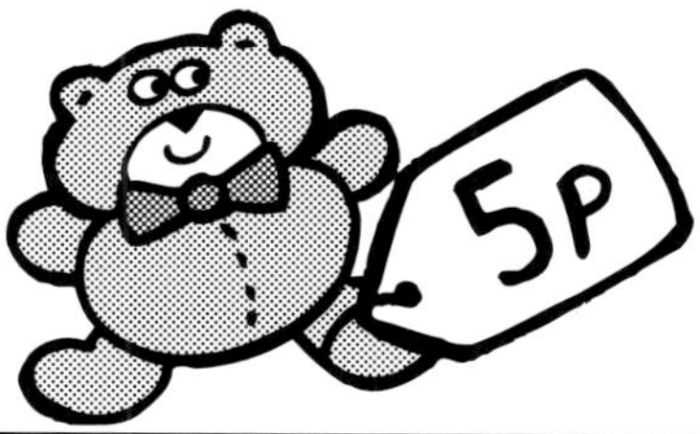

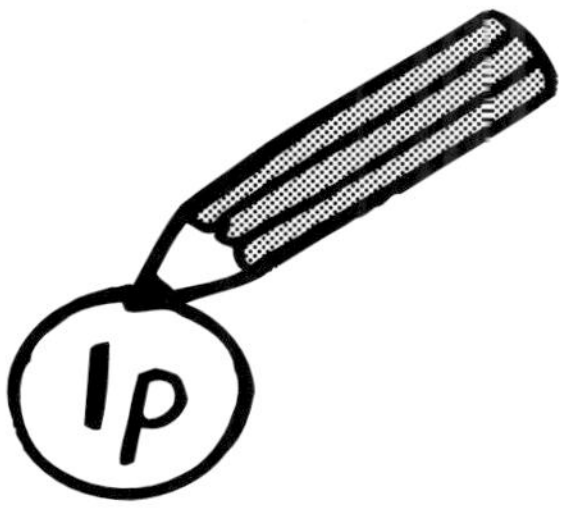

p

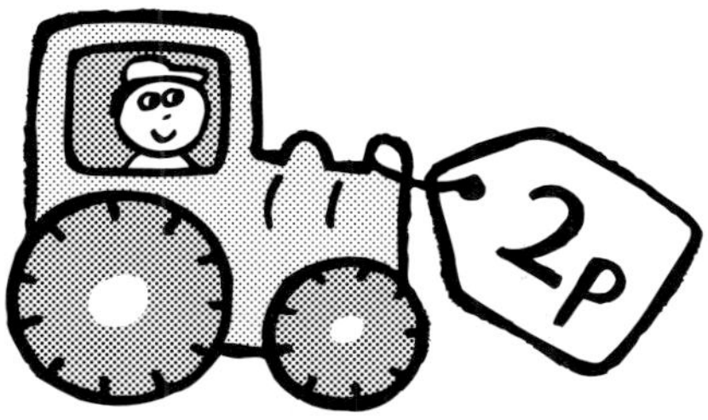

p

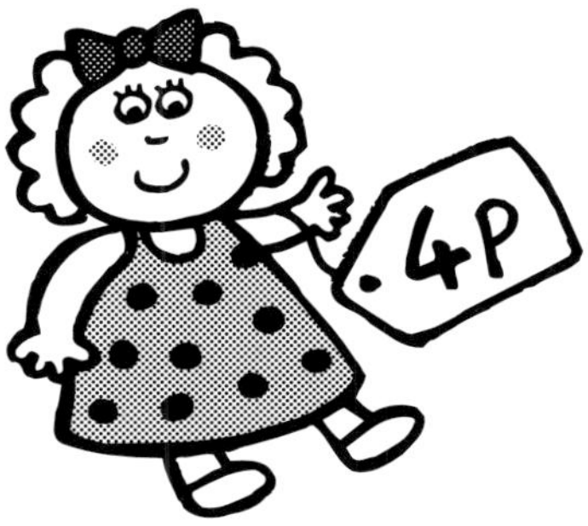

p

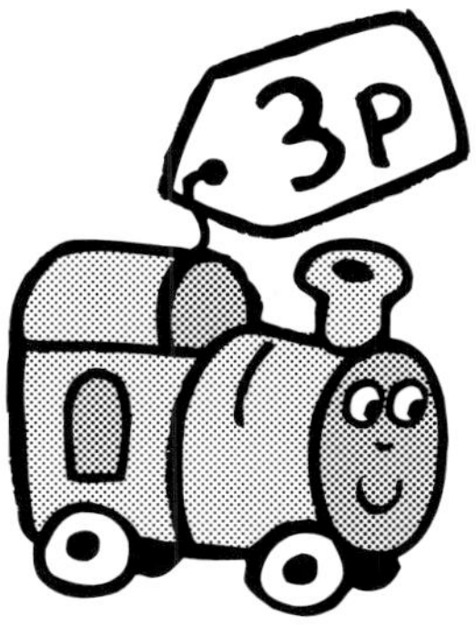

p

4	Before you start IP 5	● Can count with 1p coins. Notes/date:

Name . . . . Class . . . .

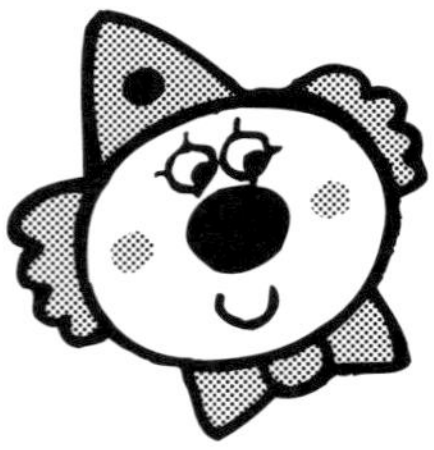

How many

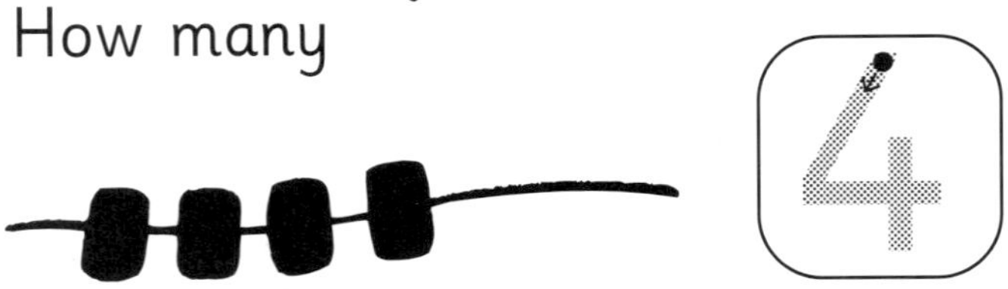

4

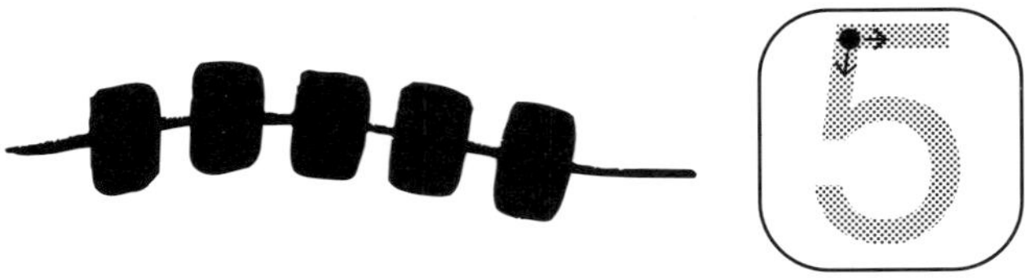

5

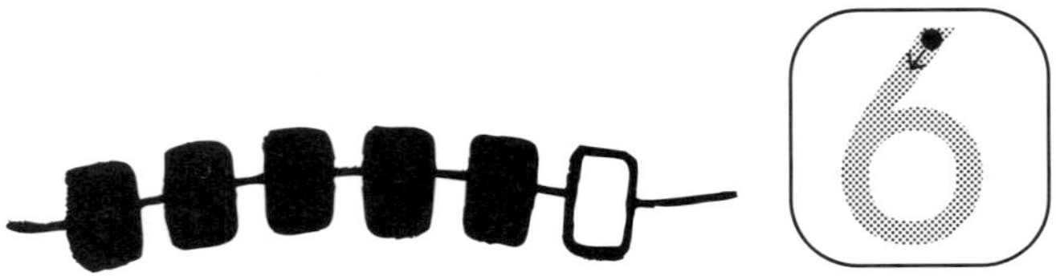

6

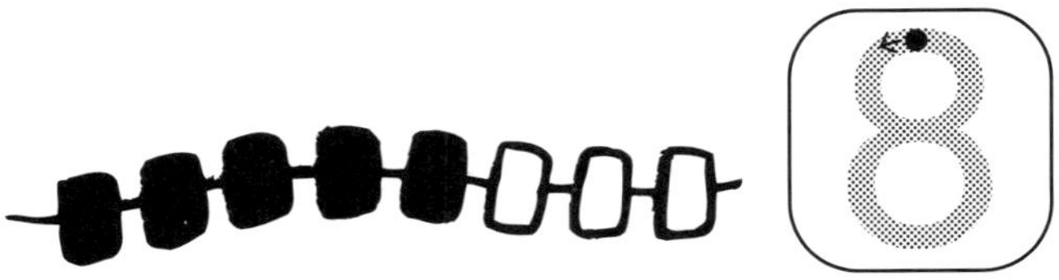

8

Draw 7

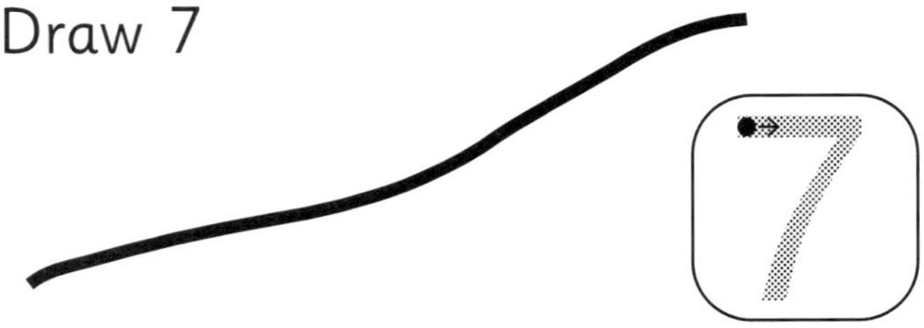

7

Draw 9

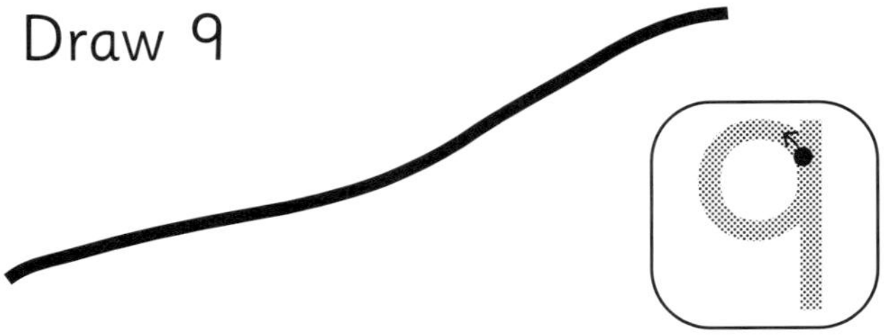

9

Draw more

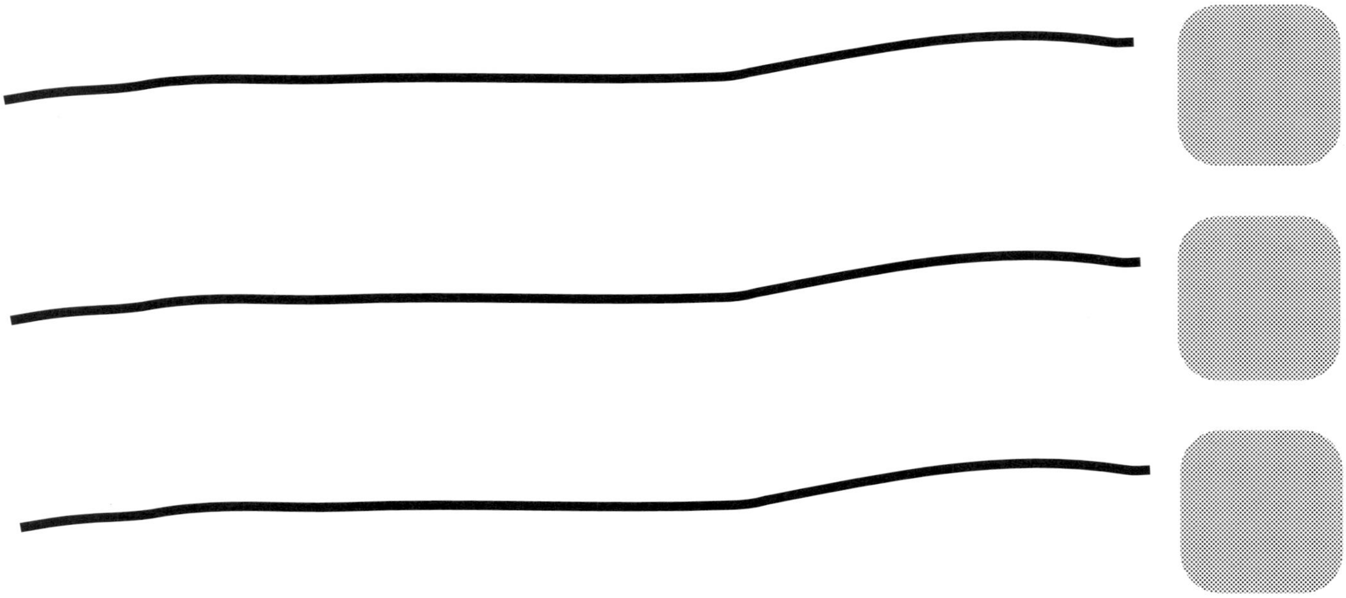

Can visualise numbers as 5 and a bit more.
Notes/date:

Before you start
IP 14

5

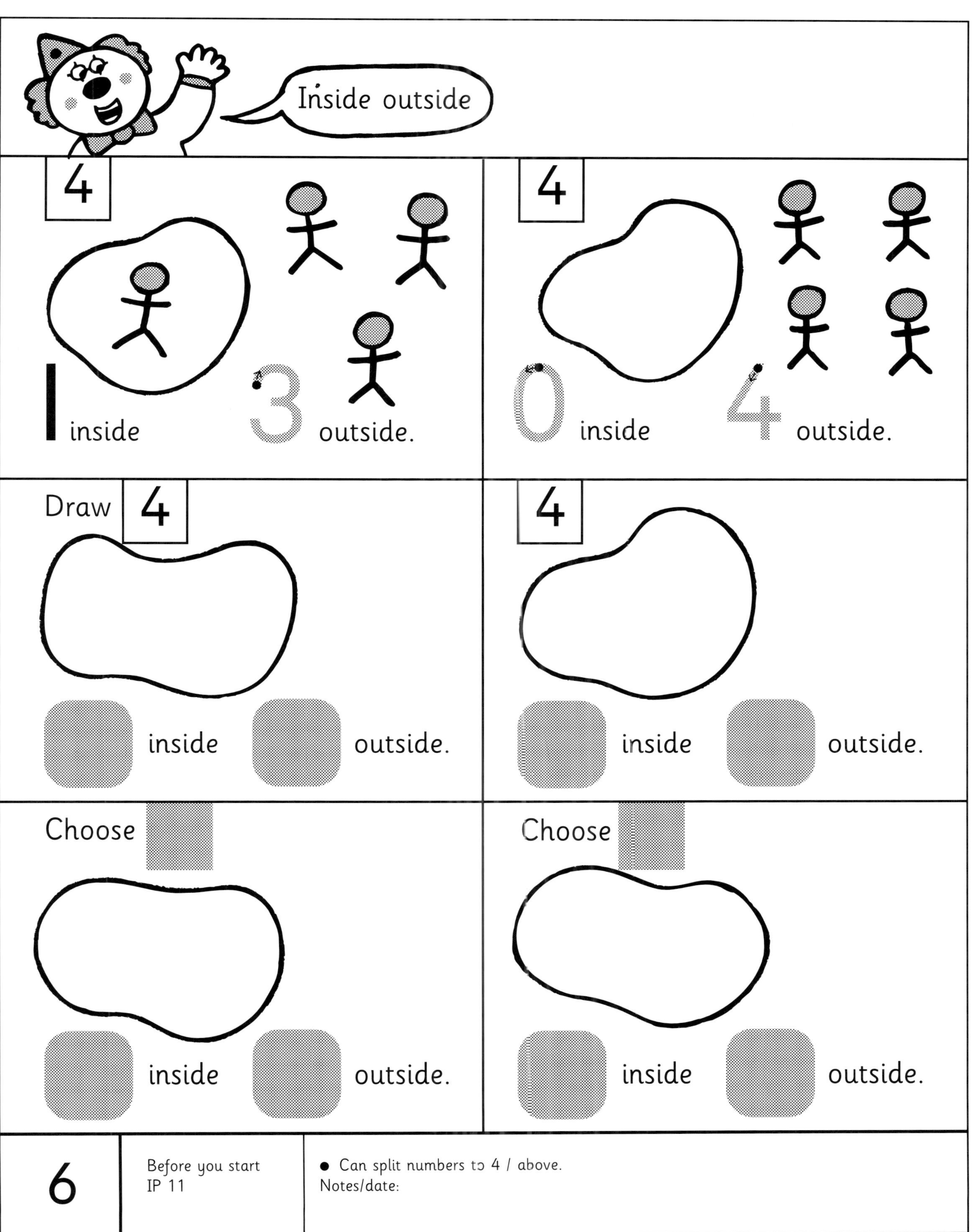
Inside outside
4
I inside   3 outside.
4
0 inside   4 outside.
Draw 4
inside   outside.
4
inside   outside.
Choose
inside   outside.
Choose
inside   outside.
6
Before you start
IP 11
● Can split numbers to 4 / above.
Notes/date:

Start with **5**

Draw what you did

4   1

• Can split numbers to 5 / above.
Notes/date:

Before you start
IP 13
CG Start with...

7

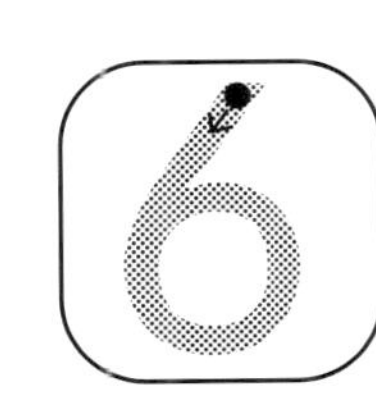

Draw what you did.

8	Before you start IP 13	● Can split numbers to 6 / above. Notes/date:

Write in the missing numbers.

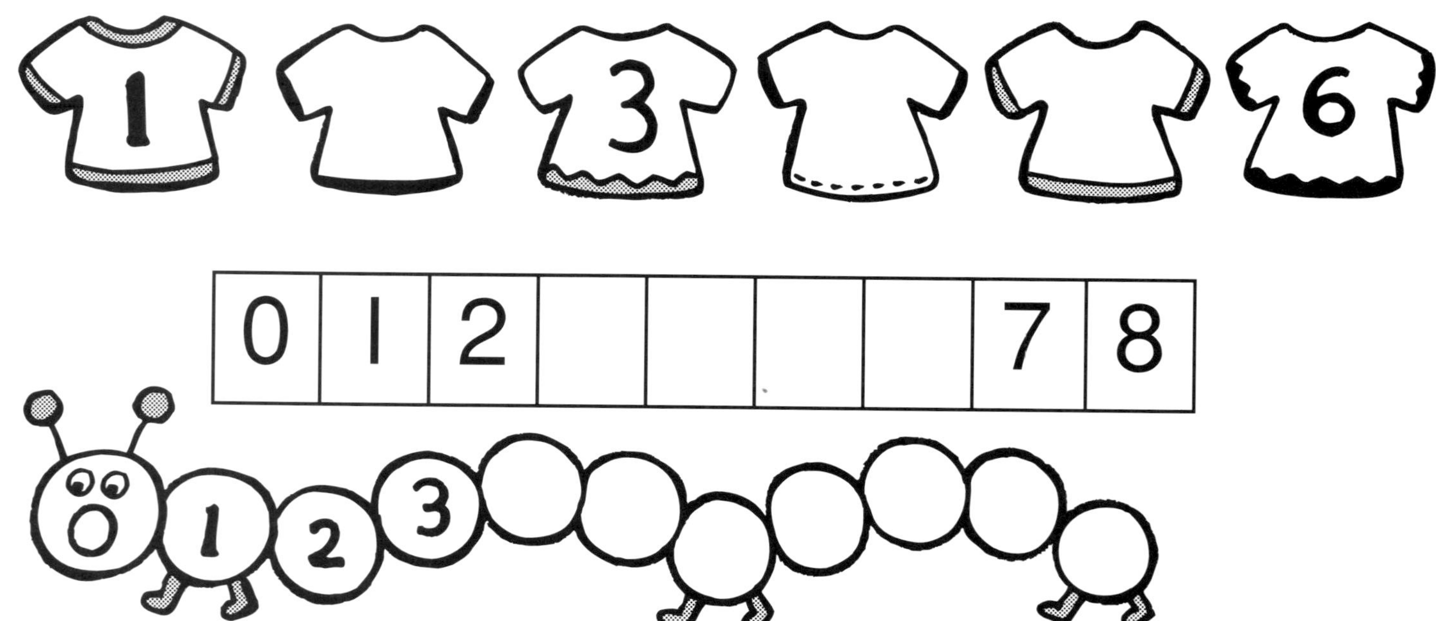

You need cards 0 – 10.
Draw

● Can sequence numbers to 10/above.
Notes/date:

Before you start
IP 15

9

42

10	Before you start IP 14	● Can hop forward on the number line. Notes/date:

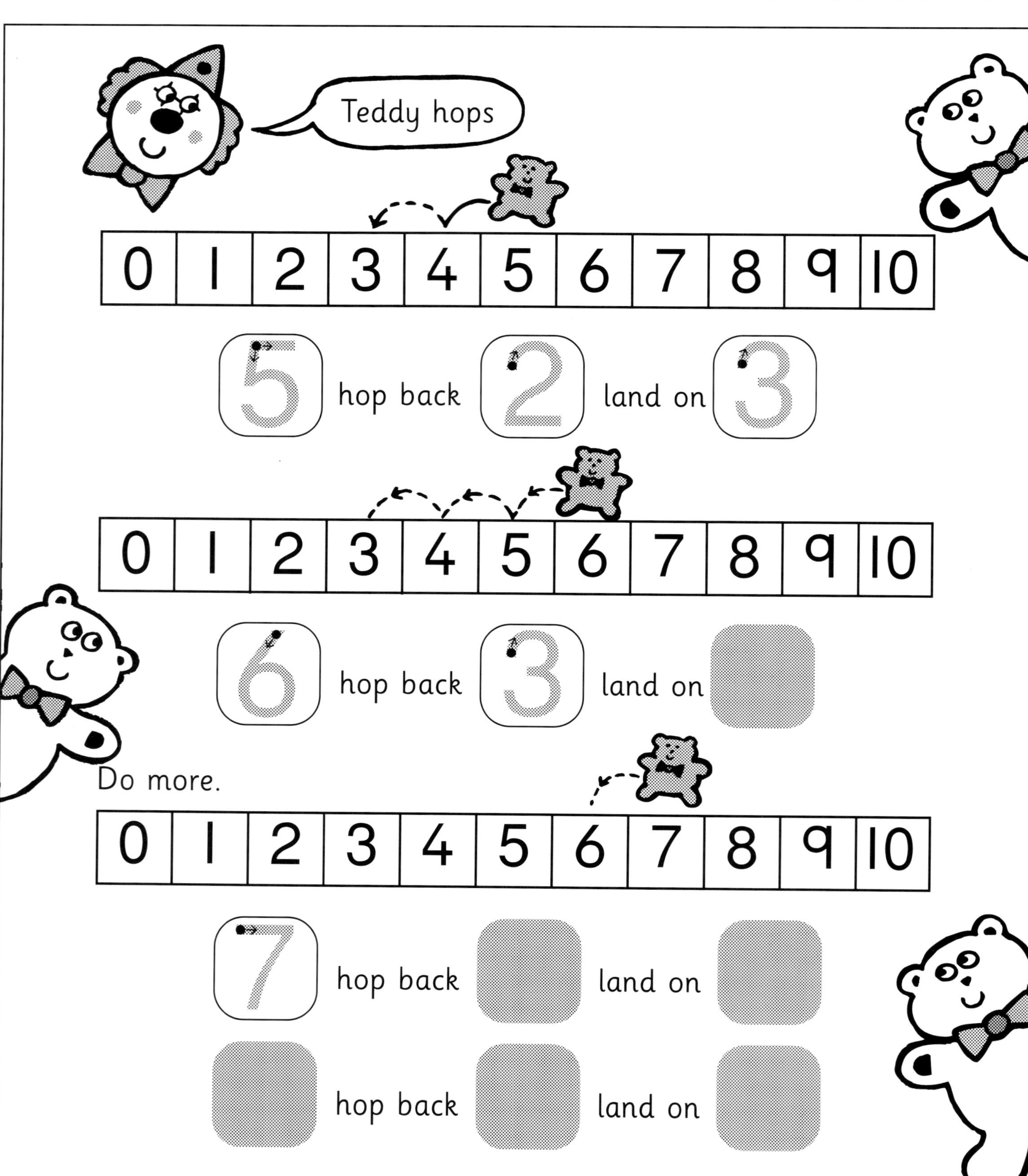

Teddy hops
0 1 2 3 4 5 6 7 8 9 10
5 hop back 2 land on 3
0 1 2 3 4 5 6 7 8 9 10
6 hop back 3 land on
Do more.
0 1 2 3 4 5 6 7 8 9 10
7 hop back land on
hop back land on

Draw more.

| 12 | Before you start
IP 11
CG Lily pads | ● Can use language of addit.on.
Notes/date: |

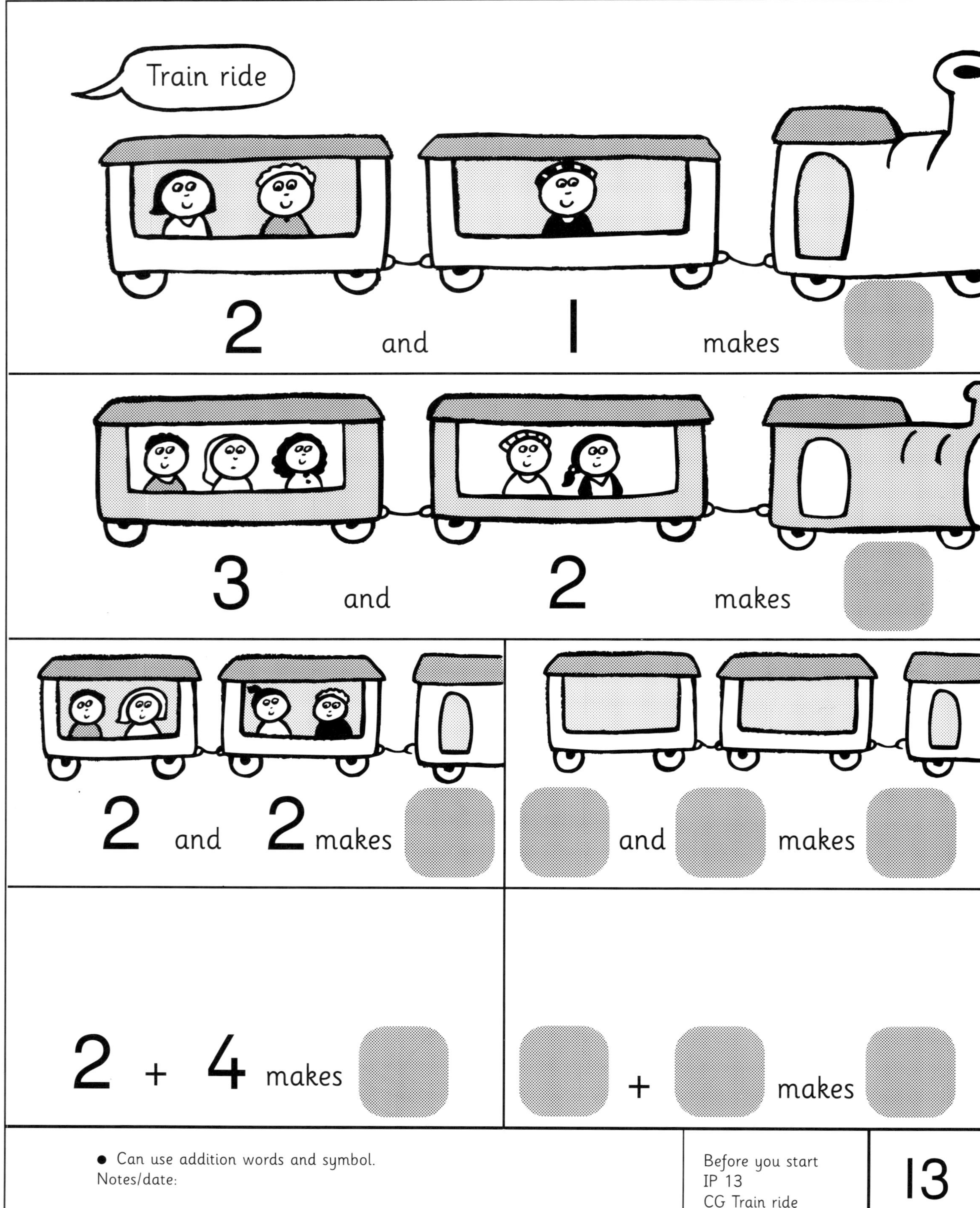

● Can use addition words and symbol.
Notes/date:

Before you start
IP 13
CG Train ride

13

0 1 2 3 4 5 6 7 8 9 10

2 and 2 makes

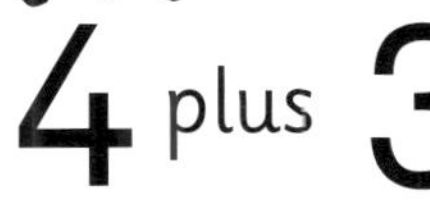 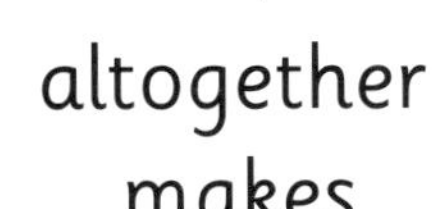

4 plus 3 altogether makes

3 and 4 equals

4 + 6 makes

Do more. Use cards 2 5 8 = +

14	Before you start IP 11	● Can add to 10 / above. Notes/date:

● Can add to 6 with + symbol.
Notes/date:

Before you start
IP 11

15

Snake game

You need
a partner
a spinner
a counter each

8 9 10 11 7 12 6 13 5 14 4 15 3 2 1 0 start

on 1 back 1 on 2 on 3

Name . . . . Class . . . .

Name . . . . **Class** . . . .

I have played these games.

Snake game

I can count to

I am going to try to

1. ____________________________

2. ____________________________

I can ____________________________

____

date __________

Snake game

Rules
- Put your counters on the tail of the snake.
- Take turns to spin the spinner.
- Move that number of spaces.
- The winner is the first to get to 15.

How many?

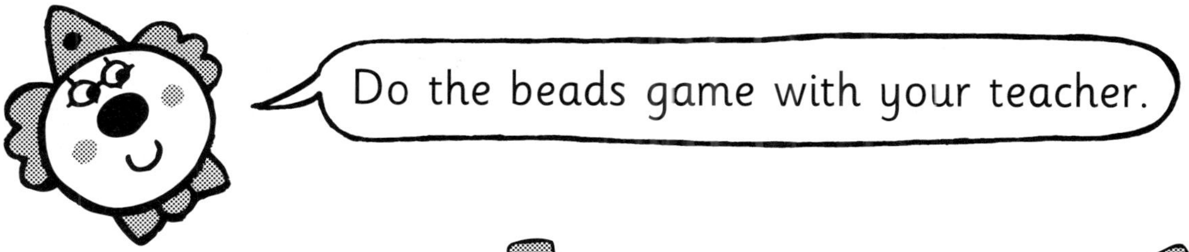

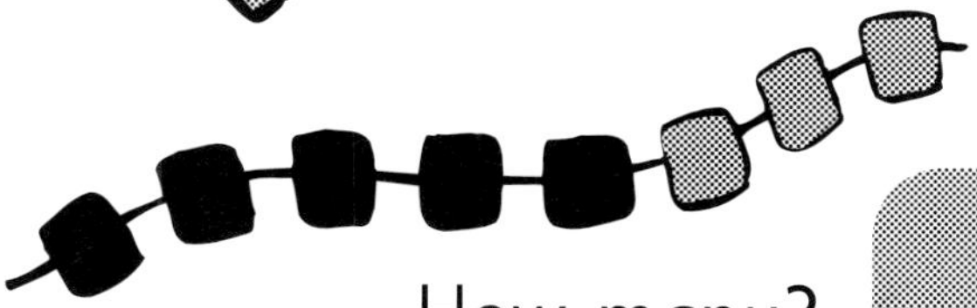

How many?

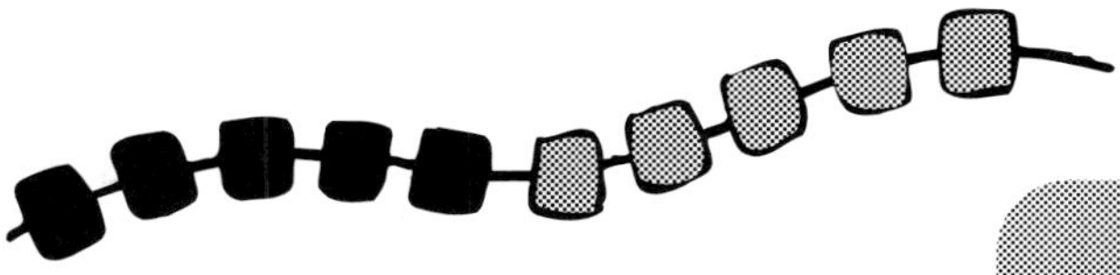

How many?

How many?

Draw

Draw

Put in the numbers.

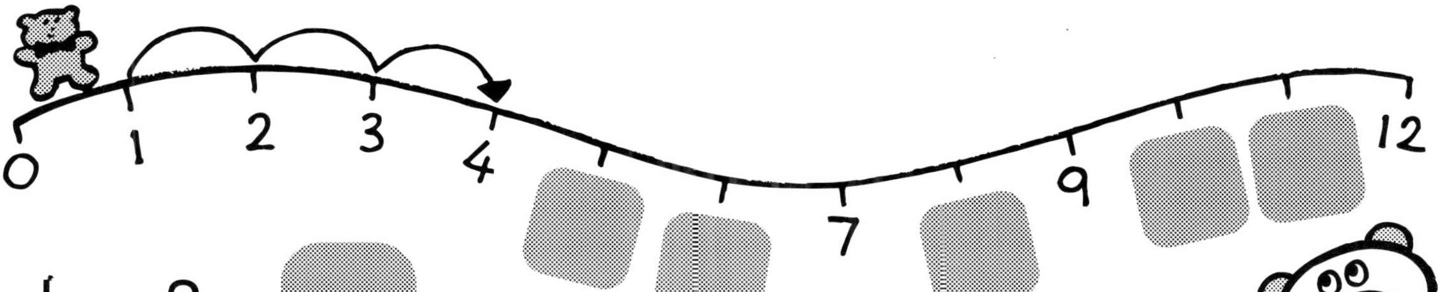

0 1 2 3 4 7 9 12

1 add 3 is

Do more.

| 2 | Before you start
IP 14 | ● Can visualise numbers in 5s and a bit more.
● Can add on the number line.
Notes/date: |

Name . . . .   Class . . . .

● Can count to 20.
Notes/date:

Before you start
IP 8

3

52

© Cambridge University Press 1998

4	Before you start IP 8	● Can write numbers to 20 / __ . Notes/date:

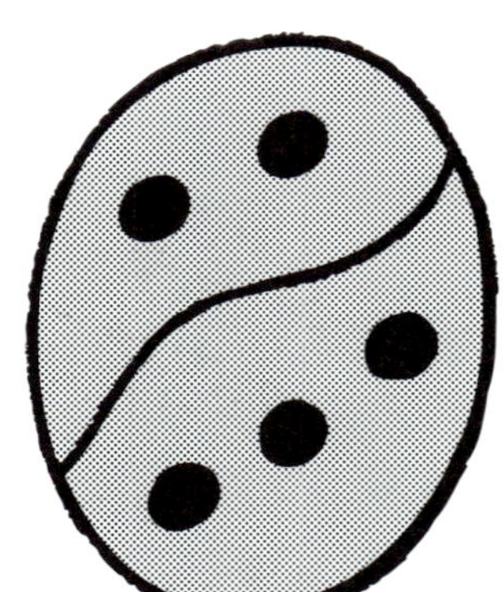

2 and 3 is 5

2 + 3 = 5

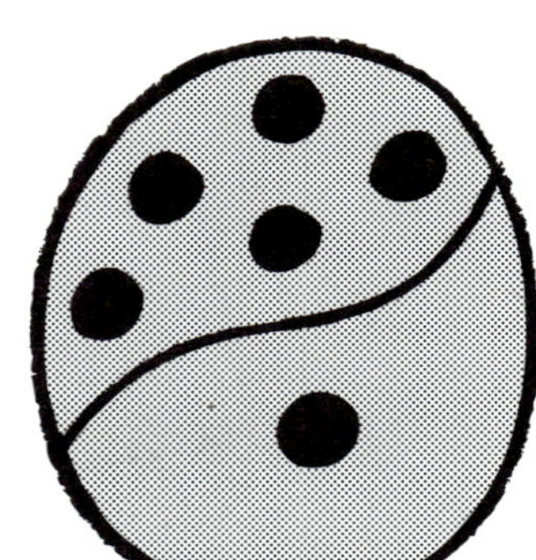

5 and 1 is 6

5 + 1 = 6

Choose numbers.

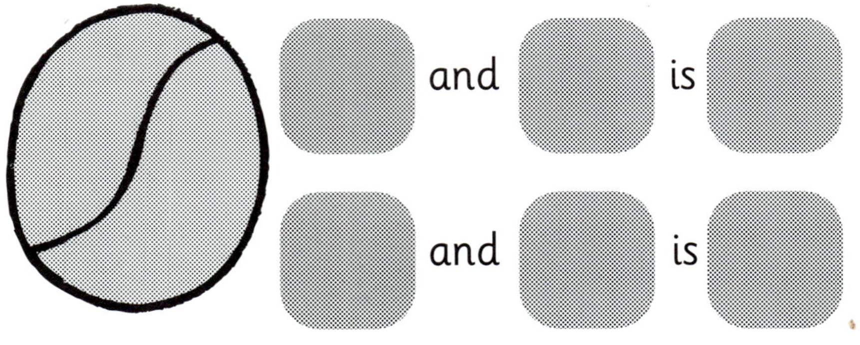

and ___ is ___

and ___ is ___

2 + 0 =

2 + 1 =

2 + 2 =

2 + 3 =

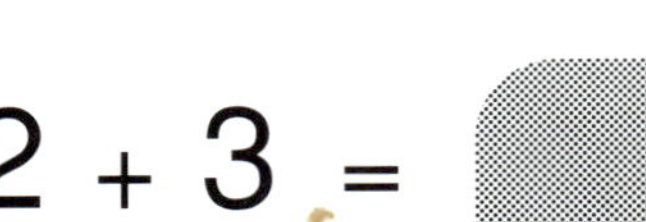

2 + 4 =

2 + 5 =

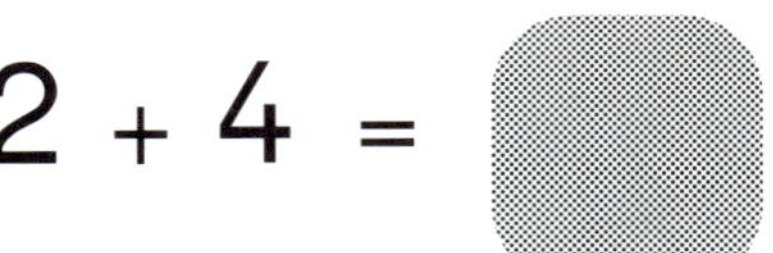

2 + ___ = ___

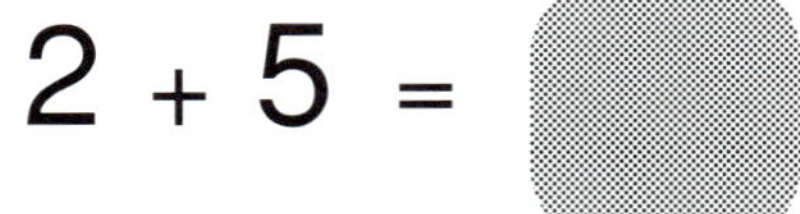

2 + ___ = ___

- Can split numbers to 10 / above.
- Can add pairs of numbers to 10 / above.
Notes/date:

Before you start
IP 5, 13
CG Splitting numbers

5

You need: cards 0–10   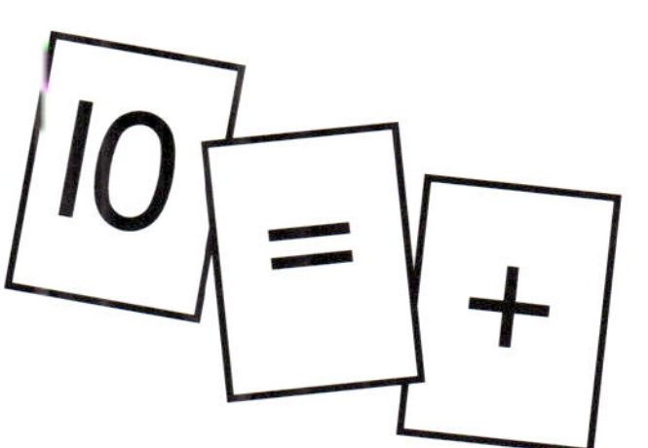 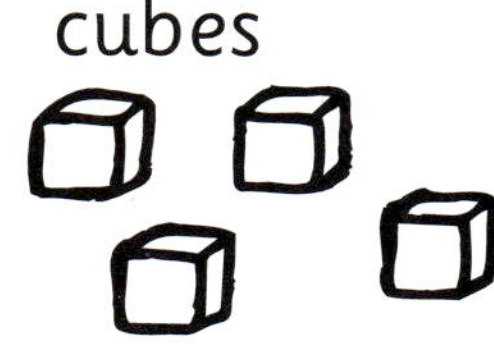 cubes

Make sums

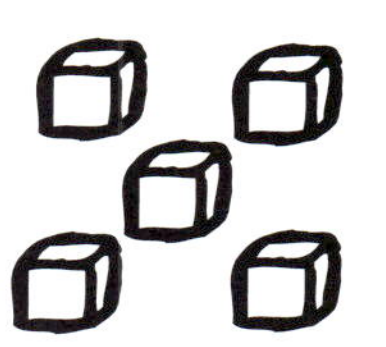 and  more   5 + 1 =
and

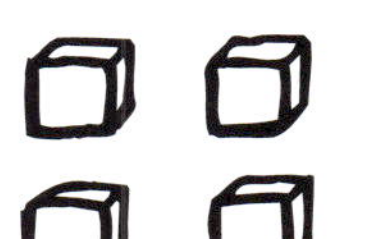 and  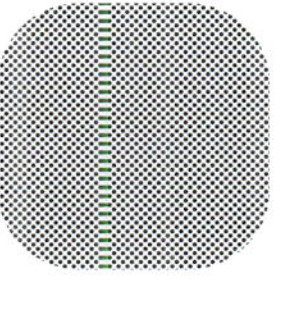 more   3 + =
add

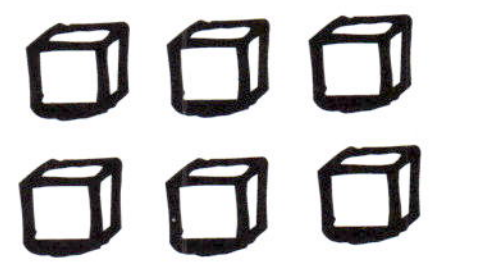 and  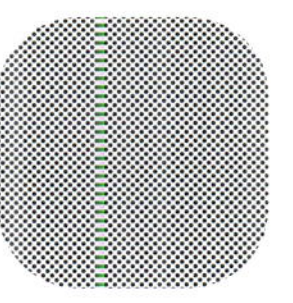 more

 and   more

Do more sums

6	Before you start IP 11, 13	● Can add with cubes to __. Notes/date:

You need: cards 0–10 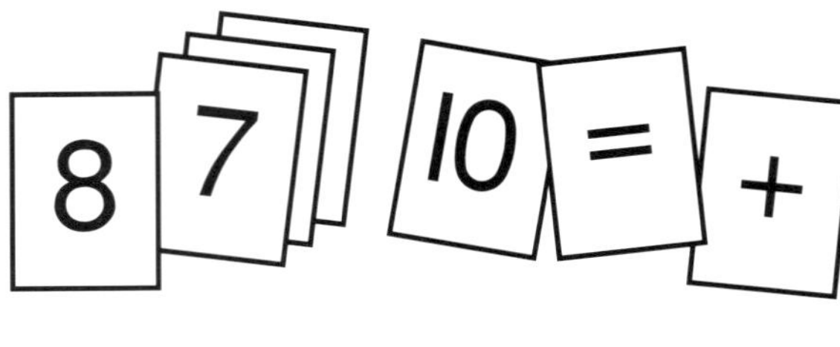 cubes 

3 + 2 = ⬚

1 + ⬚ = ⬚

2 + ⬚ = ⬚

⬚ + ⬚ = ⬚

Choose numbers 

- Can add to __ with fingers/cubes.
- Can add mentally to __ .

Notes/date:

Before you start
IP 13

7

add 1 to a number

1 add 1 → 2
2 add 1 → 3
3 add 1 →
4 add 1 →
5 add 1 →

add 2 to a number

1 add 2 →

2 add 2 →

3 add 2 →

add 2 →

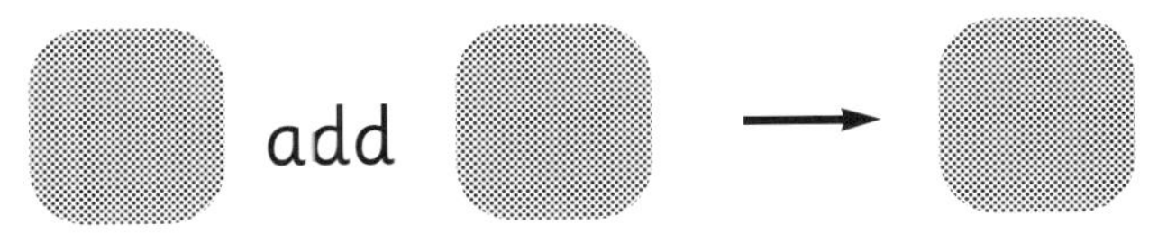
add 2 →

add ☐ to a number

1 add ☐ →
2 add ☐ →
3 add ☐ →

Try more

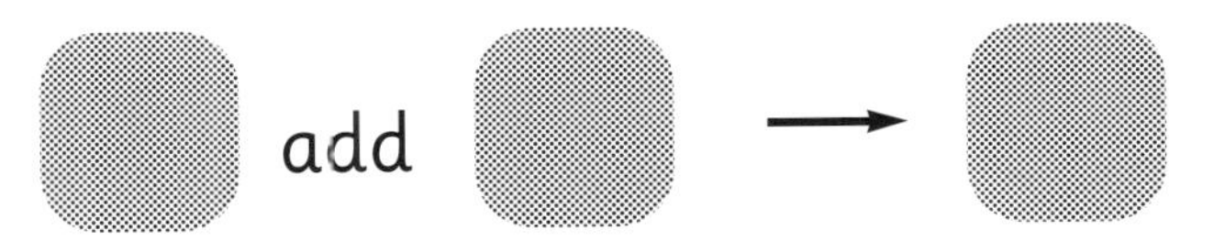

add →

add →

add →

8	Before you start IP 15, 13 CG In your head	● Can add 1, 2, __ mentally to numbers up to __ . Notes/date:

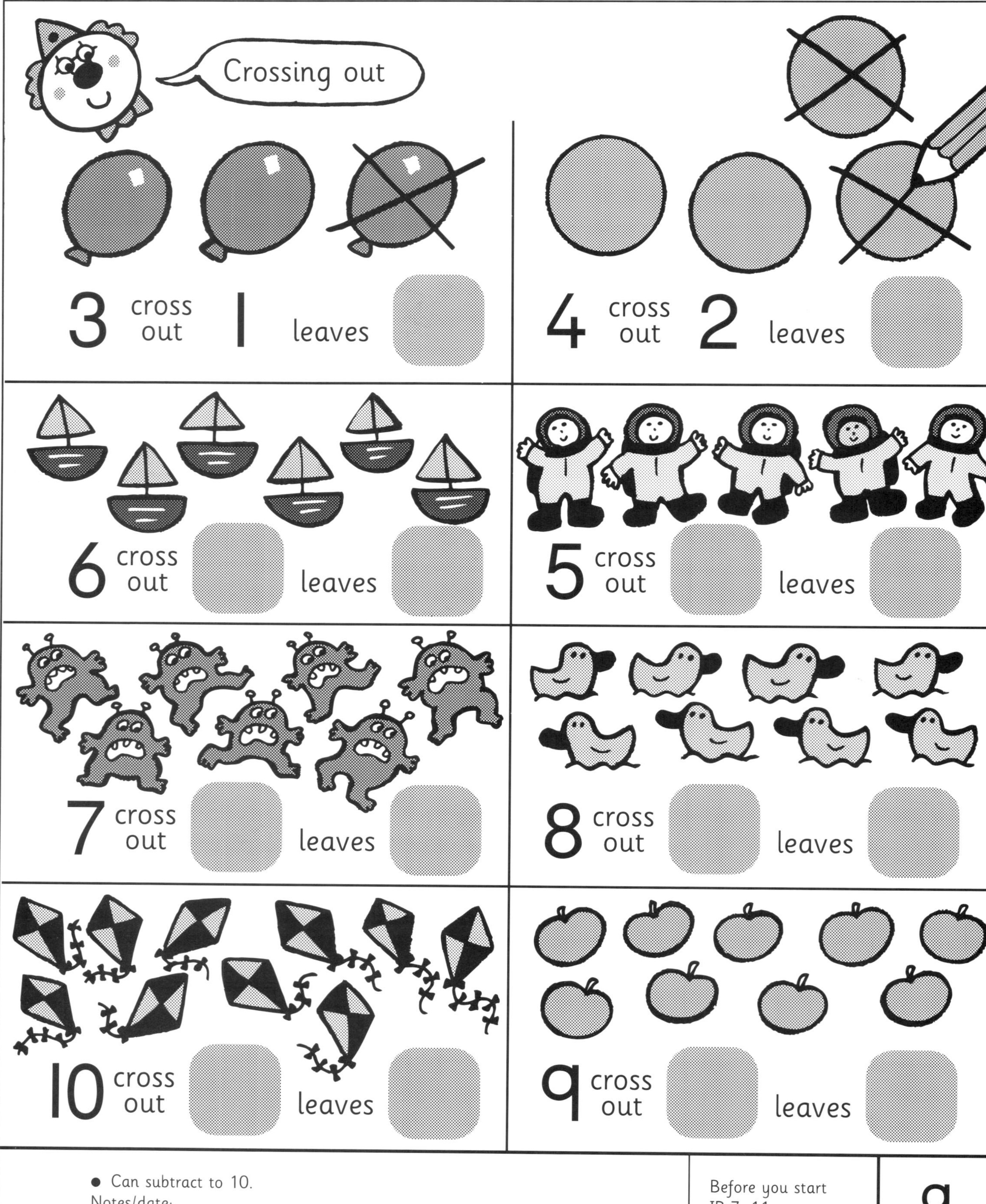

● Can subtract to 10.
Notes/date:

Before you start
IP 7, 11
CG Crossing out

q

$5 - 3 = 2$

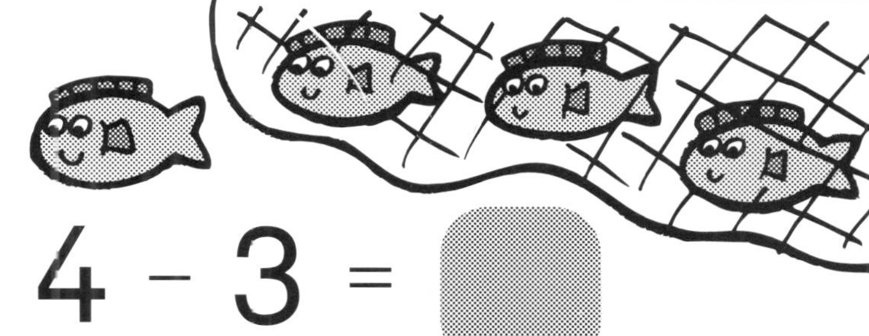

$4 - 3 = $

$5 - 2 = $

$7 - 2 = $

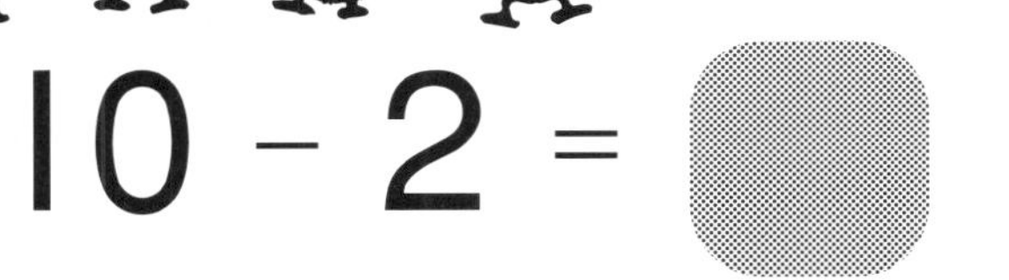

$10 - 2 = $

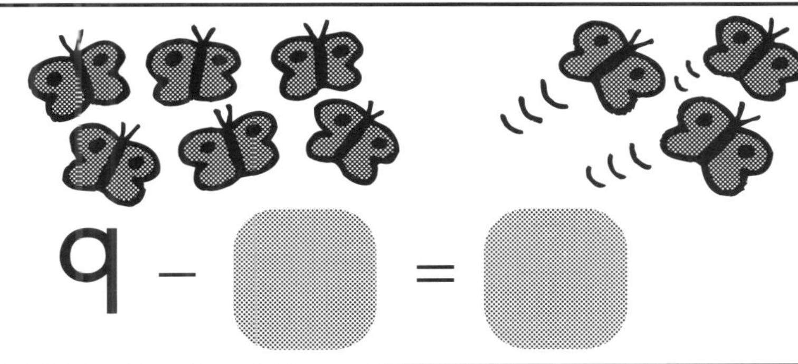

$9 - = $

Cross out

Draw 

4 cross out 1 leaves

cross out leaves

Draw more

10	Before you start IP 11	● Can subtract from numbers to __ . Notes/date:

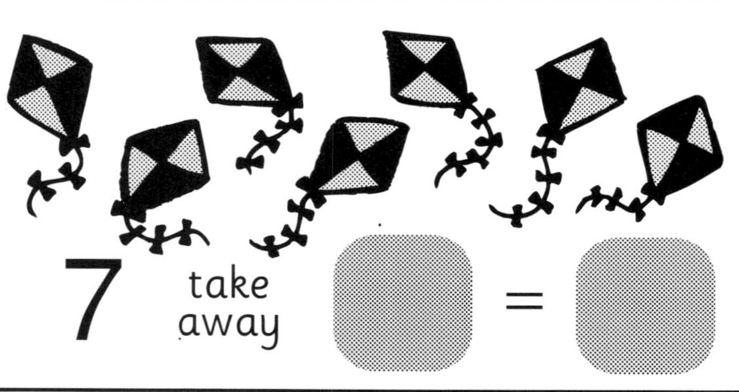

	Before you start	
● Can subtract from numbers to 10 with fingers/cubes. ● Can subtract mentally from numbers to __ . Notes/date:	IP 11 CG Take away	**11**

© Cambridge University Press 1998

Name . . . .   Class . . . .

Start with 10 mice.
Draw what you did.

| 12 | Before you start
IP 13 | ● Can record number bonds to 10.
Notes/date: |

Make all the dominoes up to 10.

  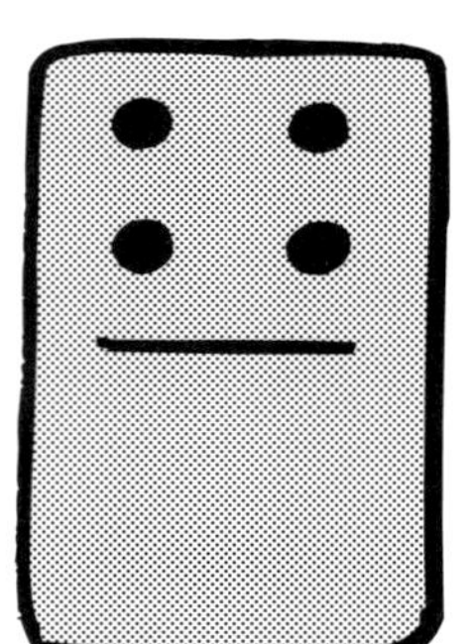

Join pairs to make 10.

 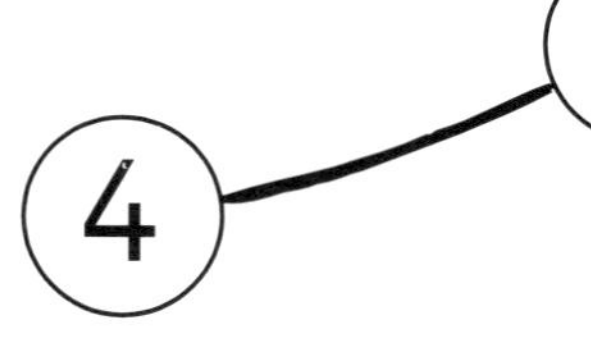 

 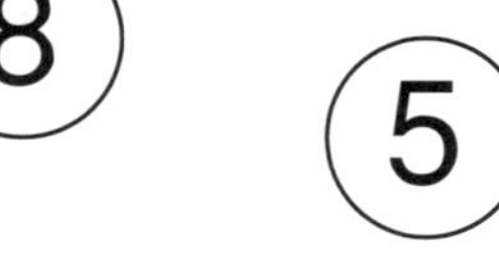 

$0 + 10 =$

$1 + 9 =$

$2 + 8 =$

$3 + 7 =$

$4 + = 10$

$5 + = 10$

$6 + = 10$

$7 + = 10$

$8 + = 10$

$9 + = 10$

$10 + = 10$

Can you see the number pattern?

● Knows number bonds to 10.
Notes/date:

Before you start
IP 7, 13
CG Make up to 10

13

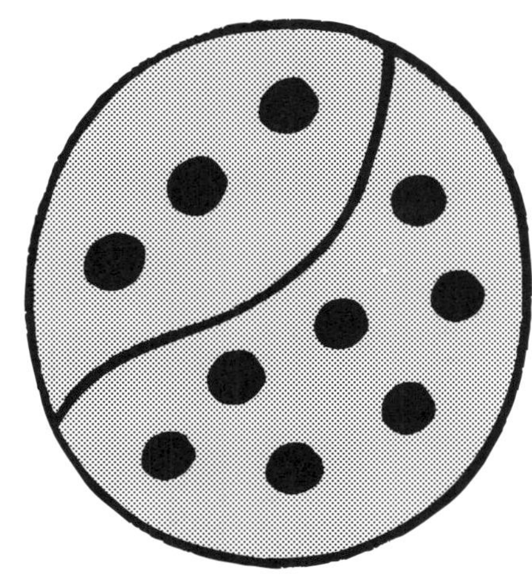

3 and 7 is 10

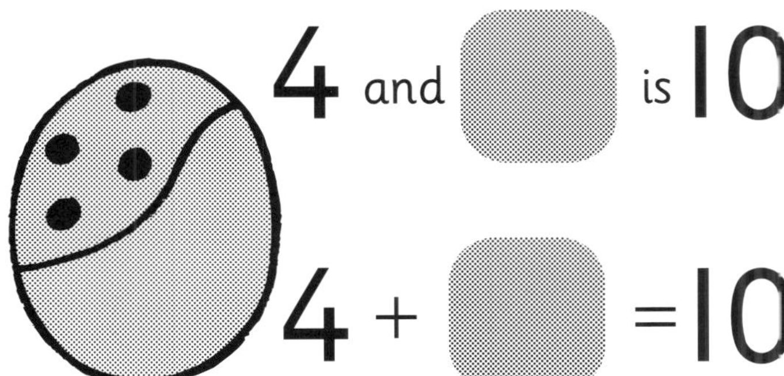

4 and ☐ is 10

4 + ☐ = 10

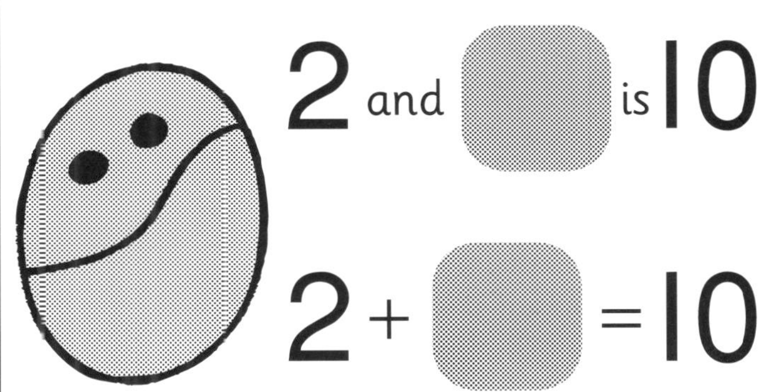

2 and ☐ is 10

2 + ☐ = 10

Choose numbers

☐ and ☐ is ☐

☐ and ☐ is ☐

☐ + ☐ = ☐

☐ + ☐ = ☐

14	Before you start IP 13, 16 CG Splitting numbers	● Can split numbers to 10 / above. Notes/date:

● Can find unknown number.
Notes/date:

Before you start
IP 12

15